PERGAMON INTERNATIONAL LIBRARY
of Science, Technology, Engineering and Social Studies
The 1000-volume original paperback library in aid of education,
industrial training and the enjoyment of leisure
Publisher: Robert Maxwell, M.C.

Microelectronics and Society

For Better or for Worse

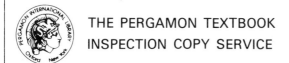

Other Club of Rome titles

BOTKIN, J. *et al.*
No Limits to Learning: Bridging the Human Gap

GABOR, D. and COLOMBO, U.
Beyond the Age of Waste, 2nd Edition

GIARINI, O.
Dialogue on Wealth and Welfare: An Alternative View of World Capital Formation

HAWRYLYSHYN, B.
Road Maps to the Future

DE MONTBRIAL, T.
Energy: the Countdown

PECCEI, A.
The Human Quality

PECCEI, A.
One Hundred Pages for the Future

A Related Journal

TECHNOLOGY IN SOCIETY*
An International Journal

Editors: G. Bugliarello and A. G. Schillinger,
Polytechnic Institute of New York

This interdisciplinary journal creates a single forum for the discussion of the political, economic and cultural roles of technology in society, social forces that shape technological decisions, and choices open to society in the use of technology. Subjects include, among others: technology assessment; science, technology, and society; management of technology; technology transfer; appropriate technology and economic development; ethical and value implications of science and technology; science and public policy; and technology forecasting.

* Free specimen copy available on request.

Microelectronics and Society

For Better or for Worse

A Report to the Club of Rome

Edited by
GÜNTER FRIEDRICHS

and

ADAM SCHAFF

PERGAMON PRESS

OXFORD · NEW YORK · TORONTO · SYDNEY · PARIS · FRANKFURT

U.K.	Pergamon Press Ltd., Headington Hill Hall, Oxford OX3 0BW, England
U.S.A.	Pergamon Press Inc., Maxwell House, Fairview Park, Elmsford, New York 10523, U.S.A.
CANADA	Pergamon Press Canada Ltd., Suite 104, 150 Consumers Rd., Willowdale, Ontario M2J 1P9, Canada
AUSTRALIA	Pergamon Press (Aust.) Pty. Ltd., P.O. Box 544, Potts Point, N.S.W. 2011, Australia
FRANCE	Pergamon Press SARL, 24 rue des Ecoles, 75240 Paris, Cedex 05, France
FEDERAL REPUBLIC OF GERMANY	Pergamon Press GmbH, 6242 Kronberg-Taunus, Hammerweg 6, Federal Republic of Germany

First edition 1982

Library of Congress Cataloging in Publication Data
Main entry under title:
Microelectronics and society.
(Pergamon international library of science, technology, engineering, and social studies)
1. Microelectronics—Social aspects—Addresses, essays, lectures. I. Friedrichs, Günter. II. Schaff, Adam. III. Club of Rome. IV. Series.
TK7874.M4795 1982 303.4'83 AACR2 81-23493

British Library Cataloguing in Publication Data
Microelectronics and society.—(Pergamon international library)
1. Microelectronics—Social aspects
I. Friedrichs, Günter II. Schaffs, Adam
III. Club of Rome
303.4'83 TK7874
ISBN 0-08-028956-8 (Hardcover)
ISBN 0-08-028955-X (Flexicover)

Printed in Great Britain by A. Wheaton & Co. Ltd., Exeter

281267

Preface

By now it is well known that the Club of Rome has taken as its central problem "the predicament of mankind" and how man can overcome this predicament which appears to be intensifying in the years ahead. All the previous reports to the Club of Rome, starting with the "Limits to Growth", have focused on this problematique. And this present book is no exception. It looks at a new phenomenon which is already having a profound effect on our lives and will have an even greater impact in the foreseeable future, namely, the new microelectronics-based technologies. Microelectronics, through miniaturisation, automation, computerisation and robotisation, will fundamentally transform our lives and impinge on most of its facets: at work, at home, in politics, in science, at war and at peace.

Recognising the importance of this change, the Club of Rome asked the editors of this book to prepare a document which will set out the challenges, opportunities and problems as clearly as possible, and enable a wide public to understand what the development of microelectronics implies for them and thus stimulate an informed public debate on what strategies to take to ensure that the new technologies will work for the benefit and not the detriment of mankind.

The subject is necessarily complex and has many aspects, so this book tries to illuminate the various facets of the problem. It starts with a general introduction which raises the major question: Are we dealing only with a new technology, or are we facing a new industrial revolution? To give a general familiarity with the history of the development of microelectronics and the basic technology which it entails, this is followed by a chapter on the actual technology. Then there is a discussion of the particular applications which already are in operation or in development. The authors of the chapters on the technology and its applications have consciously avoided speculation on what might be created in the long run, and have concentrated on the current innovations. These are already striking enough. The report has three chapters on the economic and social aspects of the use of these innovations in general and then specifically from a management and a worker's perspective. Of key importance here are such questions as the impact on employment, on how enterprises need to be managed, and what kinds of work will be generated by these changes. Next the report looks at a number of broad questions. What impact will this have on the Third World where labour-intensive industries are threatened? What impact does it have on the conduct of war, since microelectronics forms a vital core for the new instruments of war? What will be the effect on the information technologies which are at the heart of the present-day bureaucracies: does it mean greater public participation or more centralised control? And what does all this mean for global relations in a world of strong and increasing interdependence? These are key questions to which solutions must be found. And although the report does not seek to prescribe solutions, the final chapter of reflections looks at a possible way in which the central issue of the impact on work—the rationale of many societies and human lives—can perhaps be resolved.

Just from a brief review of the material, the reader can see the multiplicity of questions involved in the impact of microelectronics on society. And he or she should probably not try to read this book at first from beginning to end, but choose what topics and themes are most provocative.

*   *   *

The writing of this report to the Club of Rome lasted only eighteen months. In the very beginning, the editors felt that an international and multi-disciplinary group of authors, working from different locations, would not be able to write an integrated report. On the other hand, they did not want to publish just a collection of papers. In its present form, however, the text comes close to the demands of an integrated report, since the authors were able to meet a few times and rework their drafts after common critical discussions. But there was no attempt to standardise the chapters. The reader should take advantage of this disparity to explore the different perspectives and approaches which this book contains.

* * *

The editors and authors have many people and institutions to thank for their help and assistance. In the first place must come Dr. Aurelio Peccei, the Club of Rome's President, and the other members of the Club. It was Dr. Peccei who motivated us to write on this issue, which is one of the key developments of our time, and he encouraged us throughout our work. Both he and his colleagues realise that there are great opportunities and risks involved in the use of microelectronics and that we must immediately adapt these chances to human needs before it is too late and the possibility of beneficial adaptation has passed.

Secondly, we must thank the authors who did their very intensive jobs under the difficult conditions of only being able to meet three times. We hope that they found this project as intellectually stimulating and rewarding as we did.

Thirdly, we would like to thank Stephen C. Mills who acted as the project co-ordinator and, together with Peter Tamasi, diligently edited the text. They and their colleagues at the European Coordination Centre for Research and Documentation in Social Sciences in Vienna played an important role in keeping this decentralised enterprise together. Without this co-ordination, the book would never have been completed.

Finally, and crucially, we should like to express our warm thanks to those who contributed to the project's financing: it was their material backing which brought the project to completion. Most important was the help and friendly encouragement of the Volkswagen Foundation in Hanover, which provided the essential core funding of the project. Then three other sources enabled the group to meet each other: the Friedrich Ebert Foundation in Bonn; the Duttweiler Foundation in Zürich; and the Deutsche Gesellschaft für Internationale Entwickung in Berlin, which also financed a meeting with outside experts from different countries and continents.

In conclusion we would like to remind our readers that we strongly hope that this first compact view of the societal impact of microelectronics will start a real public debate. Even if it is critical of our report, such a debate will help to bring this crucial question to people's consciousness and can lead to the adoption of socially wise strategies.

GÜNTER FRIEDRICHS
ADAM SCHAFF

Contents

ix

List of Contributors

BARNABY, FRANK
Former Director of Stockholm International Peace Research
Institute (SIPRI), Hampshire, England

CURNOW, RAY and CURRAN, SUSAN
Probit Consultancies Ltd., Norwich, England

EVANS, JOHN
European Trade Union Institute, Brussels, Belgium

FRIEDRICHS, GÜNTER
German Metalworkers Union, Frankfurt, Federal Republic
of Germany

IDE, THOMAS RANALD
T. R. Ide Consultants Inc., Scarborough, Ontario, Canada

KING, ALEXANDER
International Federation of Institutes for Advanced Study,
Paris, France

LAMBORGHINI, BRUNO
Department of Economic Research, Olivetti Co., Ivrea, Italy

LENK, KLAUS
University of Oldenburg, Oldenburg, Federal Republic
of Germany

RADA, JUAN F.
Centre for Education in International Management, Geneva,
Switzerland

SCHAFF, ADAM
European Coordination Centre for Research
and Documentation in Social Sciences, Vienna, Austria

1

Introduction: A New Industrial Revolution or Just Another Technology?

ALEXANDER KING

What is all the fuss about microelectronics which we read about in the newspapers, and why do some people, including a number regarded as soberly reliable, suggest that it is ushering in a new industrial revolution? After all, the average person is as yet aware only of a limited number of gadgets which display directly their microelectronic characteristics—the digital watch, the pocket calculator and a flood of electronic games which appear in the shops in readiness for Christmas. He has seen, of course, advertisements for word processors, which seem to be a new and very expensive form of typewriter, but he has probably not had the opportunity yet to play with one. In addition, he is naturally aware of the mystique of the computer and realises that it is beginning to impinge on his life when he sees strange electronic lettering on his bank statement and his electricity bills and when he makes flight reservations, but these hardly seem to him or her as earthshaking innovations which will impact seriously on his or her style of life.

The last two centuries have seen an increasing flood of inventions and discoveries; these have reinforced one another and made possible new products, new production methods and new industries which could not have existed without the discoveries of the fundamental research laboratories. Indeed, the present relatively high levels of industrial productivity and of the prosperity of the industrialised countries is based on them. Technological innovation leads to an increase

1

in manpower productivity, lower prices and wage increases, which in turn produce larger markets and greater employment. As a consequence, there has been a continuous increase in the size of the world market over the years and hence of economic growth, thus belying the fears of the Luddites and Machine Wreckers of the late eighteenth century, that the replacement of men by machines would lead to widespread unemployment and misery. In reality, exactly the opposite has happened. Why then should the development of microelectronics be so different? Will it not in its turn lead to the creation not only of new wealth, but of new industries, new markets and new jobs as well? What has made such an august body as the National Academy of Sciences of Washington state in a recent report[1] that "the modern era of electronics has ushered in a second industrial revolution . . ., its impact on society could be even greater than that of the original industrial revolution"? It is the purpose of this book to examine this exciting development, with all its promise for the betterment of society as well as its dangers and to attempt to give at least some indications of the issues at stake.

In reality, the penetration of microelectronics into economy and society is already insiduously extensive, but not always visible to the public—robots in the workplace, not yet very many and the majority in Japan; automation of clerical functions; electronic banking systems; a vast and sophisticated array of controls of military equipment, defensive and offensive; new domestic consumer goods; and, of course, computers everywhere. If there is indeed to be a new revolution of economy and society, its agents have already established cells in all the sectors of economy of society; they are infiltrating, quietly as yet, throughout the whole fabric of society. It is easy to see that a great new information industry has been born and that it will have a significant impact on our way of life, but there have been other dramatic developments of technology in the past—air travel, nuclear bombs and power; and perhaps there will also be tremendous changes due to the application of biological discoveries—for instance through genetic engineering. The question is whether the difference between

---

[1] *Microstructure Science; Engineering and Technology,* National Academy of Sciences, Washington, 1979.

these and microelectronics is merely qualitative or of a different order of magnitude.

Before we describe a few of the immediately significant applications of microelectronics, it may be useful to say a few words about the history of the development. Until about thirty years ago, electronic devices including the early computers depended on the use of vacuum tubes or valves which were relatively bulky components and relatively power-hungry. Then came the transistor, a device using the properties of semiconductor materials, a product of solid-state physics. These generally consist today of silicon, a non-metallic element, very plentiful in the earth in the form of its inert oxide silica or quartz, to which minute quantities of impurities such as boron or phosphorus are added in discrete regions, altering its electrical properties. Semiconductors act as minute electronic switches, barring the flow of electrons or allowing them to pass. The transistor, being much smaller than the vacuum tube, quickly replaced it in all electronic equipment, with the consequence that the new generation of electronic equipment such as computers, television and radio sets became much more compact and, of course, it produced the doubtful blessing of the transistor radio. However, transistors and other components still had to be wired together, and as a single piece of equipment might have thousands of such components which had to be connected, electronic manufacture still remained a complicated and costly operation and the products were still relatively bulky. The further miniaturisation of such circuits came from the defence technology requirements of the United States, since weight and volume reduction were necessary for their development plans. Defence research contracts given to industrial research laboratories led to a breakthrough in 1959 with the development of the concept of the integrated circuit. Hitherto transistors were produced in batches as thin wafers of silicon semiconductor which were then cut into separate units, later to be wired together with other components. In the integrated circuit the transistors were wired together while still in the silicon sliver, at first using wires as such, but rapidly advancing to a process in which they were replaced by minute aluminium conductors deposited *in situ* on the silicon. These were the first integrated circuits, initially relatively simple, but as techniques were developed they rapidly became smaller and much more complex. Today the most closely packed integrated circuits, containing not only transistors but also other components such as resistors and diodes, may contain nearly 100,000 components in a chip measuring 5 millimetres across and with the aluminium connections about thirty times thinner than a human hair. As stated in the *Worldwatch Paper* on microelectronics,[2] "in three decades, a whole roomful of vacuum tubes and other components has been reduced to the size of a cornflake". And the process of miniaturisation is by no means finished; it is confidently asserted that by the end of the present decade, chips containing at least a million elements will be available.

A further consequence of this amazing development is that through the devising of mass production methods, the cost has nose-dived to the extent that whereas in 1960 a single transistor cost about $10, today a transistor embedded in an integrated circuit costs a fraction of 1 cent.

Further developments enabled the production of special forms of integrated circuit in which the central processing unit of a computer is put on a silicon chip and can be programmed to carry out a complex series of functions. This is the microprocessor, which is essentially a minute, silicon-chip-carried device to process information which is fed to it.

Thus the central feature of the microelectronic revolution, if revolution it is, is the minute silicon chip microprocessor. It means, in fact, that a brain and a memory can be provided in an unbelievably small size for any piece of equipment devised by man and at a very modest cost.

The present consequences of this remarkable development can be demonstrated vividly by computer development. The first electronic digital computers, introduced at the end of the Second World War, were bulky installations containing as many as 7500 relays and switches, 18,000 vacuum tubes and 7,000,000 resistors. The transistor generation was considerably reduced in bulk and today's silicon chip generation is 300,000 times smaller, 10,000 times faster, much more efficient in its use of power and at the same time much more reliable.

---

[2] Microelectronics at work: productivity and jobs in the world economy, *Worldwatch Paper 39,* Washington, October 1980.

This trend will continue until home computers are within the means of almost everyone.

The microprocessor development is significant in two senses. First, it has given rise to a new, important and exceedingly fast-growing industry at a time of economic recession. Secondly, its products can be applied in nearly all sectors of society and economy, with great potential for industrial productivity increase, for the reduction of dull and dirty work and for the creation of new wealth. It is the second of these which we shall be dealing with mainly in this book.

Amongst the applications of silicon chip technology, already in development, the following are listed to give an indication of the wide range of possibilities which have emerged even at this early stage (many of these are described in greater detail in later chapters):

the electronic watch and calculator;
the personal microcomputer;
improved functioning of the internal combustion engine;
increased fuel efficiency;
domestic appliances of many kinds, such as programmed washing machines and dishwashers, sewing machines and, eventually, the domestic robot;
information selection and retrieval;
automatic translation and interpretation;
novel traffic control systems;
new systems of public transportation;
computer aided design;
multi-purpose computer controlled machine tools;
central control of large industrial systems (oil refinery, chemical plants and steel works);
the automated factory;
the automated office;
new systems of banking, transfer of money, insurance, etc.;
environmental monitoring;
optimisation of agricultural yields resulting from computerised analysis of factors influencing growth;
electronic mail;
new information and communication systems;

voice recognition and synthesis;
the tele-video conference;
medical diagnosis and prosthesis;
computer aided educational systems.

It should be noted here that many of these applications involve also other new technological developments in addition to microelectronics; techniques such as holography, satellite utilisation and glass-fibre optics articulate well with microelectronics.

In this introductory chapter we shall restrict ourselves to looking at the impact of the integrated circuit on a few discrete situations.

Microelectronics in the Home

Microelectronically controlled devices are already in increased use in domestic appliances. Thus the majority of microwave ovens on sale in the United States are already equipped with microelectronic timing devices, and similar circuitry is beginning to be incorporated in washing machines, replacing some mechanical and electromechanical controls, more prone to disruption. Sewing machines offer still greater opportunities because of the versatility of the microprocessor as compared with traditional mechanical methods, while the reduction in the number of moving parts provides considerable economy in manufacture. In one sewing machine a single integrated circuit controls the stitch pattern in place of some 350 mechanical components in a former model. Thus the incorporation of microprocessors into conventional household equipment has great advantages of flexibility and reliability as well of making possible a greater variety of tasks and functions.

However, such innovations are hardly revolutionary, and the resulting improvements with regard to performance, economy or diminished need for maintenance will not always be recognised as due to microelectronics. Many more striking innovations are, however, in store and will spread gradually over the next few decades. Take the home computer, for instance. Not so long ago, one had to reckon on a hundred thousand dollars or more for the very cheapest computer; today, compact home computers can be bought for $1000 and the price de-escalation is by no means at the flattening-out point. Furthermore, miniaturisation through the integrated circuit provides it as a neat device which rests comfortably on a desk. At present, such instruments for home use are on the desks of a few enthusiasts who use them for calculations, local problem-solving and communication with colleagues. Soon, however, the utility will increase greatly as its costs decrease, new links to other systems are made and external constraints diminish the value of the conventional communication and information services. Thus one can envisage that the desk console of the home computer, linked to a television screen, will in the future become a basic and central utility for most households in the industrialised countries. Through this equipment the family bills will be paid, after the bank balance has been thrown on the screen, electronic mail which has been delivered at different times during the night will be read during breakfast and the various electronic newspapers and magazines to which the family subscribe will be scanned or read. The equipment will give access to data banks and to the quasi-totality of human information. It will open the way to a wide range of education possibilities and to training courses in craft pursuits and artistic activities; it will be used by the children when doing their homework. If any member of the family is engaged in research work, the computer will give access to colleagues and permit dialogues on problems. As an entertainment facility it will not only display the regular television programmes and provide a repertoire from which to select musical items, but also computer games of all sorts or chess contests with friends near or far. The housewife will, if she wishes, be able to do her shopping through the apparatus, scanning the shelves of the supermarket to choose the groceries she needs and having the cost deducted directly and visibly from the family bank account. However, it is doubtful if she will opt for such a service daily, rather than continue "live" shopping. The concentration of activities around the computer will tend to immobilise the family, just as TV does in many instances today—but much more so—and this will encourage her to seek human contact outside by shopping in the traditional manner. This picture is the vision of the "wired society", the "cashless society", but not, we must hope, "the alienated society", in which voting in elections and collective decision-making possibilities are available from the armchair at home—but not used.

The Automated Factory

The mechanisation of industrial production has proceeded gradually since the beginning of the industrial revolution and in the 1950s, with the advent of the computer, expensive as it was at that time, interest was aroused by the possibility of partial or complete automation. Despite some development of numerically controlled machine tools, the next decade saw only a slight acceleration of the automation trend and it was not until cheap and compact integrated circuits became freely available that real progress has been made. The original industrial revolution consisted essentially in the replacement of muscular work by steam and later electrically powered machines; the second is the incorporation of information and computerised intelligence into machines and production systems.

The approach to automation is following several paths simultaneously: (1) process control by means of microelectronic control of large, integrated industrial plants in continuous process industries such as oil refineries, chemical plants, pulp and paper manufacture and electric power generation; (2) the incorporation of computer intelligence through the microprocessor in individual process, transfer and assembly machines; and (3) integration of the latter towards the totally automated factory.

In the continuous process plants, large amounts of raw materials or chemical intermediates are subjected to a series of transformations in continuous flow, during which the maintenance of conditions such as temperature, pressure and rate of flow is critically important. This has for long been achieved by mechanical or macroelectronic devices, at least to a certain extent. The advent of the microprocessor, however, now allows a more complete and integrated control. Microprocessors are incorporated in the measuring instruments themselves and their variables are translated into electronic signals which are fed into a central computer able to assess the signals, adjust valves, heaters, coolers, etc., to control each of the individual parts of the production process for which it has been programmed. Moreover, the great flexibility of such control systems enables the plant operation to be modified for the production of different products or mixes of product. Such processes were already previously relatively highly automated and employing little labour, hence the improvements now sketched will have little effect on employment.

It is in the second approach to automation, in which microelectronic devices are built into the individual machines, that the greatest impact is to be expected. The computer controlled cutters, grinders, drillers, etc., of the 1950s worked in response to coded instructions from the computer and were successful mainly in repetitive operations. The introduction of the microprocessor has greatly increased the range of their versatility and enables them, by changing the instructions to the computer, to be relatively easily reprogrammed for new tasks. This makes short runs of production and even "one-off" products economically feasible and thus extends the possibilities of automatic machine tools enormously.

More dramatic has perhaps been the appearance of the industrial robot. The appearance of these machines has little resemblance to the humanised automata of Čapek's play *RUR*, which gave the word "robot" to our vocabulary, or to the "androids" of *Star Wars*. They consist mainly of computer controlled jointed arms which can hold a variety of tools such as drills, spanners, paint spray-guns or welders. These robots are thus able to extend computer-operated automation to the assembly stage of production. So far, these robots, having only a limited sensitivity, have had a somewhat restricted use in the repetitive tasks of line production. They are proving extremely valuable in the welding of car bodies, in the spraying of paints and other coatings and other operations which are unpleasant or hazardous.

A primitive robot of the present generation costs about $35,000 and is capable of being programmed for a variety of tasks. They are apparently quite economic to operate at this price, mainly because of the saving of the wages of the human workers they displace. It is estimated[3] that the average robot, working a 16-hour day, costs about $4.80 per hour, which is less than half of what is earned by an assembly line worker. It is not known precisely how many robots are already at work. The majority of the present population is said to be in Japan and their production is expected to increase considerably.

[3] Peter Marsh, *New Scientist,* 24 April 1980.

However, we are seeing only the first generation of robots which are blind and relatively insensitive. Much development work is in progress towards the creation of robots which can "see" and "feel", and prototypes already exist which are capable of seeing sufficiently to assemble small equipment such as electric motors or calculators. A significant event would be the entry of some of the electronic giants such as IBM and Texas Instruments into the design of robots. With the addition of "sight" and "touch", as well as the ever-increasing intelligence capacity which the development of integrated circuits is making possible, the present generation of robots will soon be seen as extremely primitive; the number of jobs they can perform and the intricacy of their tasks is likely to extend their utility very rapidly. "Sight" and "touch" will, of course, add to their cost considerably, but the downward trend of cost, due to increased production, will eventually apply here too and the use of advanced robots in industry will increase rapidly if social obstacles can be overcome. It is impossible to foresee at this stage the eventual impact which the sensing, seeing and intelligent robot will have on employment, but we must expect that they will progressively take over assembly tasks in many industries and thus have a considerable impact on employment.

Apart from the control of continuous flow operations, these developments towards automation are proceeding in a quite piecemeal fashion, with those elements of the total manufacturing process which are most obviously open to the application of the microprocessor being automated and the rest left to traditional methods. The logical path of development will be towards an integration of the individual tasks in the highly automated factory. This may still be a long way ahead, but the ingredients are already with us. In such a plant, computer aided design of finished products and their components would precede the computerised generation of an integrated plant, in which the central computer would control and synchronise the operation of the individual minicomputers and microprocessors required to shape and assemble the final product which would then be subjected to computerised quality control. One industry likely to adopt integrated automation of this type may well turn out to be the manufacture of integrated circuits themselves.

The Automated Office

Our present economy is constructed essentially on the basis of ever-increasing productivity of the production process, whether in agriculture or in industry, and first mechanisation and then electronic control has been outstandingly successful in achieving this end, in the process, extending markets. However, the industrial revolution has, until recently, had only a marginal influence on the tertiary or service sector. In most instances the productivity of this sector is not measured and monitored; while in sectors such as education and other public services, it is essentially unmeasurable. One element of this sector, however, is increasingly regarded as ripe for automation and hence productivity increase, namely office work in commerce, banking, insurance and the public sector, activities which have grown considerably in recent decades.

Office work is essentially the processing of information as well as its storage, retrieval and communication. This has been accomplished over the centuries on the basis of paper, pens and ink, with communication through messengers and the mail. More recently, however, the typewriter and telephone, and later the copier and telex, have made inroads upon this paper-dominated activity. At the same time, the amount of information available and data to be processed have grown enormously. It is not surprising, therefore, that the information society into which we are developing and which is made possible by microelectronics is threatening office operation with radical change. The capital *per capita* in office work is less than a tenth of that per worker in industry, but this is now changing with the shift from paper to electronics which is now possible. However, it must be noted at the outset that capital intensity in office work brought about by the introduction of word processors, facsimile machines, minicomputers, business machines of many types and computerised telephone terminals will only be economically feasible if it results, as did the progressive automation of industry, in a very considerable reduction of the work force. In industry, the labour set free by mechanisation has been absorbed by the increase in market size and the expansion of the tertiary sector. It is not so clear where large numbers of redundant office workers will find alternative employment. It must be remembered too that many commercial and office units are quite small and undercapitalised, so their automation will hardly seem desirable to their owners, at least at present cost levels of electronic office machines. It is estimated that in Britain only one office in four has as yet a photocopier, one in ten a data processor of some sort and one in forty a word processor. Indeed, there are many who argue that it may be some time before the new microelectronic devices, despite the brilliance of their performance, will prove as effective as the old traditional methods using the typewriter, carbon paper, filing cabinets and the telephone, easily understood by everyone and making possible the camaraderie of the office. Nevertheless, the incursion of the microprocessor in the office is inevitable. Earlier generations of computers, large and costly, were within the possibilities of only larger corporations; the miniaturisation and low cost of the new equipment increases the potential use enormously. It is estimated that by the middle of the present decade sales of minicomputers for office work may be approaching half a million per year which contrasts dramatically with the fact that only ten years ago the total world population of computers was under 100,000. It is highly probable that large office units, such as health and employment services, banks, insurance companies and large corporations, will automate rather rapidly to form a modern office sector.

We shall not discuss here the many new microelectronic innovations relevant to office automation; some of these will be described later; suffice it to say that the extent, variety, flexibility and interconnection of the new devices are such as to make the automated office realisable in the near future. This will greatly influence the structure and organisation of office work. In the recent past, most office machinery served to mechanise traditional routine tasks; the new microelectronic equipment will go much further in changing the nature of the tasks and in necessitating radical changes in the flow of work. The result will be not only the elimination of a high proportion of existing jobs—and this may well bear most hardly on the employment of women—but requirements for quite new skills. Management of the automated office will be something quite different from that at present and, all in all, one can foresee many difficulties of transition and the need for a basic reassessment of skills and for the educational and training facilities to produce them.

*   *   *

In the advance towards automation there are two constraints; the first is the capital intensity of the process, and there may well be considerable competition for capital in the future because of the need to transform and renew much of the existing world energy generation system as well as to provide the infrastructure for a greatly increased world population; more serious is likely to be the adjustment of society in the highly industrialised countries to the conception of "jobless growth" which we shall discuss later. However, the pressure of international competition between the major industrial countries to exploit this new wave of development is likely to force each of them to create the new generation of high-productivity industries as quickly as possible.

We return now to the somewhat theoretical question of whether we are at the beginning of a second industrial revolution or is this just one more technological development, important certainly, but one which can be absorbed smoothly into the ongoing fabric of society? Revolution or evolution: this is largely a matter of semantics. After all, the first Industrial Revolution took several decades to gather its momentum and, therefore, could hardly be regarded as revolutionary in the political sense. Nevertheless, it did mark an abrupt discontinuity in the evolution of both economy and society, a sort of mutation arising from the improvement of the steam engine at a propitious moment, which enabled it to find rapid and economic application across a wide spectrum of industrial processes. We must therefore rephrase the question: is the burgeoning upsurge of microelectronics and its applications the trigger point for another major discontinuity in economic development, and consequently of society, or is it just another point on the upswinging curve of contemporary technological development derived from scientific discovery? There are partisans of both points of view, and indeed one of the purposes of this book is to fuel the debate on this issue so that, should the revolutionary character of microelectronics prove to be a reality, sufficient understanding of its consequences may be gained to enable society to guide the development for the benefit of humankind.

We are inclined to accept that the impact of the integrated circuit *is* revolutionary. No other single invention or discovery since the steam engine has had a broad impact on all the sectors of the economy. Even the availability of electric power merely gave a further, if powerful, impulse to the process of mechanisation initiated by steam power. Nearly all the other great innovations have been sectoral or vertical in their significance in creating new products and new industries. But microelectronics is not only certain to transform many of the traditional activities of the agricultural, industrial and service sectors, but, by incorporation of brain and memory as well as brawn into the new machines and systems, it will change the nature and direction of development; the first Industrial Revolution enormously enhanced the puny muscular power of man and animals in production; the second will similarly extend human mental capacity to a degree which we can hardly envisage now.

There are also subsidiary reasons for the outstanding importance of this development and of its probable social consequences. These reside mainly in the timing of its appearance. We appear to have entered a period of economic difficulty, with prospects of slow and uncertain growth which followed the unprecedented growth and expansion of the world markets of the fifties and sixties. Furthermore, the industrialised countries are troubled by rising costs of oil, a sense of vulnerability with regard to the security of supplies of energy and minerals on which their prosperity depends, coming as they do from distant regions over which they have no control, and by concern with environmental degradation. Into this scene now erupts the microelectronics potential. As an industry, it is not hungry for materials or energy and it is clean; more importantly, it is growing rapidly. Few will believe seriously that microelectronics, together with other high-technology developments, can bring back the good old days of rapid economic growth and market expansion which generated a high level of demand for goods and services and made possible full employment. It is certainly no panacea for the present economic malaise, but it is one of the few bright spots of the economy and offers good prospects of a reasonably short-term nature. Hence all the advanced industrialised countries are seizing upon the promise of this industry, pushing technological development and hoping for a share in the market for the products of this new and growing industry, or fearing that by neglecting to do so they may be opting out of the race for industrial survival. Still more important, of course, are the prospects arising from the applications of the microprocessor which may evolve relatively slowly, but which may build up towards the full industrial revolution. These prospects exist not only for the advanced, science-based industries, but by applying the new techniques to traditional manufactures such as textiles or automobiles, these industries should be able to enter the advanced sector with new vigour and competitive edge.

The Contemporary Perspective

Before proceeding to the discussion of the probable consequences of the applications of microelectronics, it is necessary to consider a number of other features of contemporary development, within which this new technology will operate. The impact of microelectronics on economy and society is indeed likely to be considerable, but it is only one of several marked and continuing trends in the evolution of contemporary society and must be considered together with these.

One of the central concerns of the Club of Rome is the importance of securing greater understanding of the interactions between the various problems and trends which constitute the complex tangle of issues which the Club terms "the world problematique". To look at the individual difficulties and new lines of progress one at a time and in isolation—which the vertical, sectoral approaches of nearly all governments to national policy make inevitable—gives a limited and often misleading understanding of the inherently complex situation. Microelectronics is almost certainly the most vigorous and all-pervasive amongst present technological developments and although it has important features which distinguish it from all other major technological advances, it has to be seen in relationship to other developments and, more importantly, in the light of conventional thinking on technological development in general.

The Political Consequences of Technological Development

Throughout human history technological advances have had a dominant influence on the development of society. The shaping of primitive tools and the discovery of the use of fire were the earliest technological achievements of *Homo sapiens,* and the domestication of animals and the discoveries which made possible a settled agriculture produced the first great transformation of society. Later, metallurgical discoveries, which made possible first the Bronze and then the Iron Age, paved the way for new types of civilisation and caused redistributions of power. However, the political impact of technology has been most clearly manifest through its military and economic effects. The replacement of the long-bow by the cross-bow was described in a contemporary papal encyclical in doomsday terms which fit perfectly with the fears which followed the explosion of the first nuclear bomb. Following the first military use of gunpowder, an accelerating stream of advances in military technology provided dominant, if often temporary, advantages to their possessors and struck fear in the hearts of their enemies, modifying the global distribution of power. The military applications of microelectronics have already, as we shall see later, proved to be an important factor in advanced military technology and will continue to be so.

An important aspect of the economic impact of technological development is the demonstrable influence which it exerts on the international division of labour. Steam-driven machines for spinning and weaving introduced by the Industrial Revolution rendered the manually operated textile industries of Asian countries uneconomic in world markets; the discovery of aniline dyes brought an end to the indigo industry of India; the Haber process for the production of chemical fertilisers ruined the Chilean saltpetre industry and, more recently, the microelectronic digital watch has eroded the Swiss industry in favour of those of Japan and the United States.

Yet, despite such influences and the arising of great new industries such as those of chemicals, pharmaceuticals and electronics, which simply could not have appeared without the discoveries of the fundamental research laboratories, most economists have tended to underestimate or to ignore the role of science as an autonomous force in industrial and economic development, in making possible quite new technologies. It has often been assumed that technological development arises from the interaction of economic forces and hence that new processes and products appear at the appropriate time as part of the intervention of the "invisible hand". A corollary of this is the "technological fix" approach which relies on technological solutions for economic problems. There has been a certain degree of validity in such thinking in the past, when technological innovation resulted more often from invention than from scientific discovery. The situation has, however, changed radically in that most of today's new technology arises from the systematic research programmes of industry, backed up by "oriented" fundamental research, supported mainly by governments to provide new knowledge which, directly or indirectly, will make technological advance more sure. The exploitation of such discoveries depends, of course, on propitious economic conditions, but is only indirectly caused by them. Many contemporary technologies, such as nuclear power generation, lasers and glass-fibre devices, cryogenics and, above all, microelectronics, are of this type and their economic and political consequences have to be seen differently than in the past. Indeed, in times of rapid economic change the relatively long lead-time from laboratory to production on a significant scale means that technologies have to be planned well in advance of their significant economic impact. The tempo of science is widely different from the tempo of politics. There is simply insufficient time for the economic forces to operate in the creation of new technologies and too great a reliance on the technological fix can be dangerous.

From past experience we must assume, therefore, that further technological developments will continue to have major political consequences, especially through their military applications and by shifts in the pattern of industrialisation and of trade. Following the successful impact of scientific and technological innovations on the outcome of the Second World War, the postwar decades have seen an extraordinarily large increase in the resources of all the developed countries devoted to scientific research and technological development. Indeed, by far the greatest proportion of such activity in human history as a whole has taken place during that period and, with the long lead-time from discovery to production just described, we have as yet seen only the beginning of its harvest. However, well over 90 per cent of world research and possibly as much as 95 per cent is undertaken in the relatively few highly industrialised countries, with only marginal efforts in the developing nations, thus tending to reinforce the economic superiority of the former. Deep consideration must therefore be given to the consequences of current technological development, of which that of microelectronics is probably the most significant, but by no means the only major item, on the North-South issue. In view of the growing appreciation of the importance of the interdependence between nations, including the vulnerability of many industrialised countries to the disruption of external supplies, which recent petroleum crises have highlighted, this matter is of key importance to countries of all categories.

In recent decades the overwhelming proportion of national expenditure on research and development in the industrialised countries has been justified by three objectives: defence, increased economic growth and national prestige; only relatively marginal funds have been allocated for the attainment of social objectives and rather little thought has been given in advance to the social, cultural and political consequences of technological development. The appearance of unwanted and unforeseen side-effects, such as those of environmental deterioration, loss of work-satisfaction and alienation of individuals from society, have, of late, given rise to much *post facto* concern and it is clear that there is great need to foresee the socio-cultural consequences of the next rush of technological development and not wait until they overwhelm our societies.

The Spectrum of Development

The heterogeneity of the distribution of resources over the planet is one of its most striking characteristics. Sources of energy, mineral deposits, soil propitious for agriculture, climatic conditions and many other features are spread very unevenly and greatly influence the degree and nature of development of different regions. This is further complicated by the fact that population distribution, for historical and cultural reasons, by no means coincides with resource distribution. Thus there is very great disparity between the levels of development of different localities which cannot be correlated with any single natural factor.

It has become conventional to discuss countries as "developed" or "developing", and this is the basis of the so-called North-South dialogue. Generalisations based on this vague and oversimplified concept can, however, be quite misleading when applied to issues such as the consequences of future technological development, since neither developed nor developing groups are anything like homogeneous. In reality there is a graded spectrum of national situations depending on present economic performance, availability of natural resources, soil quality and agricultural potential, climate, environment, capital accumulation, levels of education and human skills, technological and managerial capacity, national and political will, cultural determinants and a whole series of other elements.

At one end of the spectrum we have a few countries such as the United States, the Soviet Union, Canada, Australia and South Africa, well endowed in minerals and energy resources, with accumulations of capital as well as of technical and managerial skills, leading producers of food and with relatively stable political systems. Then follow the other industrialised countries of the world, including those of Europe, East and West, and Japan, also rich in capital and skills, but lacking natural resources. As we continue along the spectrum, we come next to countries possessing enormous possibilities in their mineral or energy endowment, but otherwise only partially developed; these include the Middle East oil-producers and Venezuela and Mexico, while further along the spectrum are mineral-producers such as Zambia and Zaire with their copper, potentially rich, but only at the beginning of their development. A few countries of these groups, such as Brazil and Mexico, are quickly developing into rich and well-endowed nations. Beyond this are countries such as those of the Indian sub-continent and Indonesia with huge populations, some natural resources and considerable agricultural possibilities and with strong, educated, intellectual élites, but still plagued by poverty and even hunger. At the far end of the spectrum, in different degrees of development and natural endowment, we come finally to the poorest of the poor, with only modest prospects of improvement.

The classification of the countries of our planet into the traditional three worlds of the market economy, the state economy and the Third World of the developing countries has proved useful for political purposes as the brittle cohesion of "the Group of 77" has shown, but it is quite inadequate for the discussion of future development and the penetration of the new technologies. A much more fine-structured approach is required. For example, amongst the developed countries, those of Europe and Japan—with their present dependence on external supplies of energy and raw materials—will have a greater incentive than the United States or the Soviet Union to concentrate on technological innovation and to maximise the value added to their imported materials and thus they may be forced to seek new patterns of alliance with other and resource-rich regions to provide a political rationalisation of their evident interdependence.

Within the so-called "Third World" of less developed countries, the differences of condition are even greater, and this is particularly obvious with regard to their capacity to assimilate new technologies. In discussing the impact of microelectronics on different types of society, generalisations such as that implicit in the term "Third World" can be utterly misleading, and we shall have to distinguish, for example, between the potentialities of the rich oil-producing countries, nations such as India or Brazil which possess skills and infrastructures, the newly industrialising countries of South-East Asia and others which are at a much earlier stage of development.

The Coming Decades of Transition

The need to consider the new technological developments in the context of major world trends has already been stressed. It is necessary to look not only at how microelectronics is likely to mould future societies, but at the same time to consider how it will interact with other major agents of change. Looking then at some of the major trends in contemporary society which are likely to persist, it can be suggested that we are entering a period of deep transition which may last from thirty to fifty years before leading to a completely different type of world society with much greater numbers, changed values, new political and administrative structures, entirely novel forms of institutional behaviour and a technological basis very different from what we are familiar with today, which will influence lifestyles fundamentally in all nations and all cultures.

This transition is being forced upon society by a whole range of causes, of which we can consider here only three, which, may, however, be the most dominant. These are the virtually inevitable large increase in the total population of the world, barring nuclear disasters or natural catastrophe on an unprecedented scale; the need for a more or less complete renewal of the planetary energy production system; the impact of the new technologies and especially those based on microelectronics. It is doubtful that these major movements, with all the changes of power structure which they involve, can be coped with effectively with present structures, national and international, with the existing economic system or, indeed, with the traditional political ideologies.

Here we can give but a brief indication of the probable significance of each of these trends. With regard to population, there has recently been a marked reduction in the *rate* of increase of the world population. Even with the declining fertility rates, however, the absolute size of the world population will rise substantially during the next half-century and, by the year 2000 will be about 6.2 billion (U.N. median variant) as compared with 4.4 billion in 1980. In terms of sheer numbers, population will be growing faster in 2000 than it is today, with 100 million people added each year as compared with 75 million in 1975. About 90 per cent of the growth will occur in the poorest countries. It is also estimated that the world population will have reached 9.4 billion by 2025, before levelling off at about 10 and 12 billion by 2074. These figures are, of course, very uncertain for the later years, assuming, as they do, no abrupt discontinuity in present trends, resulting from dramatically changed attitudes or major catastrophes.

The magnitude of the change will vary greatly from region to region. The U.N. assessment indicates that while the total population will have increased by rather more than 40 per cent by the end of the century, that of Africa will have grown by 75 per cent, Latin America by 65 per cent, South Asia by 55 per cent, East Asia by 24 per cent, the United States and the Soviet Union by 17-18 per cent each and Europe only by 7 per cent. By the end of the century then, the proportion of the total population of the world residing in the present industrialised countries will be about 20 per cent only, unless there should be mass immigration in the meantime, while this may fall as low as 15 per cent by 2050. Sheer pressure of numbers in the Third World will considerably alter the North-South relationship, and this realisation should encourage some countries to work towards some new International Order.

At the end of the century, the age composition of the population of different countries will vary greatly. In many countries of Africa and Latin America, half the population will be aged 15 or under, thus throwing an immense burden on the education system. In the developing countries as a whole, increase in the numbers of the active work force (taken as including all the 15-64 age group) will be relatively much greater than that of the total population, while in a few countries such as Mexico, Pakistan, Bangladesh and Brazil the number of extra individuals looking for jobs will be extremely high. Such large increases in the potential work force represents, of course, an enormous growth of the productive potential of the countries concerned; however, many of these countries already suffer high rates of unemployment and underemployment and hence, after the provision of adequate food and water, the creation of jobs is likely to be a main priority. This situation has, of course, grave significance with regard to the choices to be made within the industrial development policy and the use of labour-intensive technologies would seem to be essential. However, much of the technology transferred from the industrialised countries is capital intensive and provides few jobs in relation to the capital employed. This situation will undoubtedly influence the rate of introduction of microelectronics and automated industries.

The second main factor of the transition is the need for a more or less complete renewal of the world energy supply system as petroleum becomes scarcer and more costly. This fundamental transformation will be concurrent with the large increase in world population and will likewise make high demands on capital. The new energy system will compromise a wider mix of energy sources than hitherto, with an initially gradual, but later accelerating trend towards the use of renewable resources. There are great uncertainties with regard to the development of many of the non-traditional sources of energy on a significant scale. Fusion energy holds great promise, but is unlikely to be available on a large scale within the next half-century. The various means of supplying solar energy are likely to be extremely important in the long term, and the large solar radiation capture in most tropical countries may indicate them as future main suppliers of energy. In the meantime, bioresource utilisation could be of great use in many developing countries during the transition. The big question in the world energy situation is whether research, development, capital accumulation and construction can be completed before major shortages and the high cost of traditional fuels being serious economic and social difficulties.

Microelectronics, through the new information technologies and the microprocessor, could have an important influence in reducing energy demand. These techniques are intrinsically less energy-hungry than the traditional manufactures and techniques of today. Furthermore, their use in systems control could provide major energy savings; for example, in the operation of internal combustion and other engines, in the possibility of constructing new and efficient public transportation systems and innumerable other applications. The new information devices are likely to obviate unnecessary journeys—indeed video-conferencing could be a threat to the prosperity of the airlines.

The third element of the transition period is likely to be the impact of the new technologies, especially those based on microelectronics and on genetic engineering derived from recombinant DNA. The former is the subject of this book, and at this point we stress merely the fact that its development will be taking place concurrently with the two other trends which we have mentioned, competing with their needs for massive capital inputs and interacting with them, hopefully generating wealth and bringing eventually great social benefits. The management of the transition will not be easy. Existing governmental structures, based on vertical sectoral policies and short term in their vision as a consequence of short electoral cycles, will not be adequate; the present academic and research organisation based likewise on vertical disciplinary specialisation, while necessary, is insufficient to penetrate the multifaceted problems; economic policies based on quantitative growth through the stimulation of consumption will hardly prove effective in an era of resource limitation; social philosophies which follow rather than precede technological revolution will not make for smooth transition; above all, international arrangements which do not take into account the reality of the interdependence of the nations and do not provide the necessary degree of equity can only fail. The transition is thus essentially political in its nature, although caused by other factors, and it does demand great understanding between the peoples and a high and imaginative effort of social and institutional innovation.

National Capacities for Science and Technology

In this period of transition the North-South issue is likely to be of growing importance, not only because of the population differential, but also as a consequence of the very different abilities to assimilate the new technologies. Studies by the OECD seem to indicate that, above a certain threshold of technological competence, diffusion of new techniques takes place relatively smoothly and even spontaneously across national frontiers and that technical innovation rates are not directly correlated with national research efforts, although a degree of indigenous innovation is desirable if full use is to be made of the totality of world research and development. Thus for the OECD countries there exists, as it were, a common technological system from which all benefit, operated essentially through the free market system, within which new technology and know-how are, in fact, normal items of trade and subject to the usual commercial considerations. Achievement of the threshold of technological competence is, as it were, the entry card to this system. Such competence is a complex of many factors which include a sizeable effort in research, both basic and applied, pilot plant and engineering prototype capacity and the existence of the necessary spread of skills which include those of management and marketing as well as those directly linked with science and engineering.

As we have seen, by far the greatest proportion of scientific and technological effort of the world is in the developed countries and hence, with a few exceptions such as India, Mexico and Brazil, countries of the Third World lie well below the critical threshold of technology. It is this which makes the assimilation of transferred technology and its modification to meet local needs and to use local materials difficult. Thus much of the technology transferred from industrialised to developing countries consists of packages of technique and know-how mainly through the transnational corporations and inevitably of a type appropriate to the consumption-driven economies of the North. In the absence of solid technological infrastructures in the recipient countries, such imported processes do not easily become organically rooted, although they may, under favourable conditions, contribute to the growth of the desirable indigenous capacity.

The importance of the "indigenous capacity" concept was one of the most important results of the 1979 U.N. Conference on Science and Technology for Development and is likely to become a major element of development activity of the next decade. This is not merely a question of establishing research institutes and programmes and of training the necessary skills, but of creating and implementing national policies involving governments, industry, agriculture and the university and educational system to cover, in an integrated manner, the totality of the spectrum of research, development, production, industrial planning, skill development and marketing; without this, the developing countries will have little possibility of evolving technologies directly appropriate to their own needs and will lag behind the countries of the North, becoming ever more dependent as world technology becomes more sophisticated. The integrated approach is mandatory; technology may be an autonomous force in development at all levels, but it is not the basis of an autonomous policy.

As far as the main thrust of this book is concerned, the question which arises here is whether the countries of the Third World, which have not yet been able to assimilate the possibilities provided by the first Industrial Revolution, will be able to benefit by the upsurge of a second and much more complex phase of technology—that of microelectronics.

Microelectronics; the Long-term Opportunities and Risks

Turning now from the general features of change to the specific impact of microelectronics, it must be stressed that the potential benefits which will flow from this new technology are so enormous that there will be no question of avoiding or slowing down their actualisation. It is because of the extent and depth of these long-term benefits and of the changes which they will bring to society if exploited to the general good, that it is necessary to look well beyond the present decade. If this is not done and developments are planned merely on the basis of medium-term awards and of narrow-vested interests, it is inevitable that governments will attempt to absorb social and other consequences by marginal adustment of existing social models and policies. The result would be an increase in rigidities and distortions as the century runs out, making it increasingly difficult to accomplish the fundamental transformations which the optimum application of microelectronics will suggest. The changes in the nature of society to be envisaged are radical and will permit no mere return to "normality", full employment and rapid economic growth as we have known it. The fundamental question is whether governments, with the support of an informed public opinion, will be capable of using the new possibilities of microelectronics deliberately and consciously to shape a better society, rather than passively attempt mere adjustment to its consequences as they appear, as a matter of expediency.

It can hardly be denied that technology has been largely determinative in shaping the life styles of the materially successful societies of today, but the credits and debits resulting from technological development were not foreseen—they just appeared. Indeed, it is only belatedly that we have identified the unwanted and often unexpected consequences of such development, or recognised the growing significance of many of their symptoms which hitherto seems trivial and acceptable. These include dangers to environment and climate, creeping desertification in many parts of the world, the population explosion, the difficulties of life in the large cities and their faceless suburbs, loss of satisfaction in work, alienation, crime and many other difficulties. Further and massive development of technology, and particularly the new wave of developments resulting from microelectronics, could further exacerbate these and hasten the breakdown of society. On the other hand, if seized positively and used as an agent for the design of better societies, it could have exactly the opposite effect. The decisions to be taken are extremely weighty for the human future.

Let us first look at the positive aspects. The promise of the microprocessor is that through its ubiquitous applications in the automation of industry and the tertiary sector, it is capable of increasing productivity to the extent that it should be possible to provide all the resources required by a country, including those of defence, health, education, nourishment and welfare, to provide a reasonably high material standard of living for everyone, without depleting or degrading the resources of the planet, with only a fraction of the physical work expended today. Initially this will be seen in terms of the elimination of dirty, boring, repetitive and dangerous jobs and of shorter working hours and years. Later, it could open the way to a society in which the individual would have time, resources and opportunities to achieve fulfilment through the cultivation of his or her specific interests, artistic, scientific, craft, educational, sporting or otherwise. This could lead to the virtual abolition of poverty and the tyranny of work. In fact the microprocessor could be the key to Utopia. However, as in the traditional tales, the path is barred by many obstacles, some real, others imaginary and hidden within the nature of present society.

It is as difficult at this early stage in the evolution of the microprocessor and information society to foresee the consequences to the individual and the community as it was to envisage the influence of the motor car at the beginning of the century, when its chief danger appeared to be to horses and pigs and the prospects of damaging the human brain through racing along the roads at the horrifying speed of 30 kilometres an hour. There are, however, some trends, already discernible, which require consideration. It will tend, for instance, to increase greatly the interdependence of individuals and nations through the instant availability of information. It will also make for greater complexity of institutions and societies, already so complex as to be virtually unmanageable. It could make possible a high degree of decentralisation of power and of decision-making which is highly desirable, but it could equally be used by unscrupulous leaders to augment and consolidate centralised power. The means will shortly exist for the electronic control of the activities and perhaps even the thoughts of everyone by "Big Brother" dictators and societies. The present concern for the privacy of data in a world where personal details are fully recorded in data banks whose computers talk ceaselessly to each other, is an initial and healthy recognition of this hopefully distant danger. It will contribute to the fragility and vulnerability of society. Even at present levels of technology, the smooth functioning of cities and societies depends on technical devices, and they are exceedingly sensitive to disruption, either through malfunctioning or sabotage. The consequences of power cuts in a huge city such as New York are frightening, and as human activity becomes more deeply computerised, such dangers could reach a new order of magnitude. Power stations, nuclear reactors, oil refineries, communications networks and data banks will all have their sensitive nerve centres, the destruction of which could suddenly disrupt the workings of society and which are very open to sabotage and political violence. Indeed, it could easily be that warfare in the future could develop into a struggle for or even between the computers. This could prove a compelling argument, even for those possessed of power mania, towards strong decentralisation.

Much more subtle and difficult to discern, however, are the social and individual consequences of the microelectronics revolution. In a strongly technology-based culture there is always a tendency towards a dichotomy between those who understand its nature and workings and those who merely push the buttons. Granted that it is not necessary to understand electronic theory to profit from television, but when the microprocessor has spread to make "black boxes" of nearly all the equipment and artifacts of life and the level of technical sophistication of those who invent and design them has soared beyond the comprehension of the many, we may be faced not with a two-cultures situation but one of sharp distinction between the few who know and the many who do not know. The emergence of a priesthood of technologists and technocrats is certainly undesirable. It is by no means inevitable, but if it is to be avoided, its possibility has to be foreseen.

More serious, however, is the possibility of isolation and alienation which a microelectronics culture could generate. We are already familiar with the hypnotic power of the television set and how it chains people to their chairs by the hour, persuading children to enter into the imagination of others, at the expense of their own. The cashless, jobless, information society could easily exaggerate this effect. The concentration of information inlets in a single room, the focus of communication possibilities, many of them impersonal and distant, the density of educational and entertainment channels, these and many other factors could immobilise and isolate the family from human outside contact. This could too easily lead to a creeping alienation of the individual, not the active counter-cultural withdrawal we see today, but a passive and insidious alienation with loss of human dignity and self-esteem. Put in starker terms, will the automation of a high proportion of human activities eventually lead to the automation of humankind? The answer is that it probably could, but if properly understood and prepared for, it could do precisely the opposite, and this is in effect the thesis of the final section of this chapter.

Employment, Occupation and the Renaissance of Society

Employment prospects are the crux of the matter. Our thesis here is that the influence of microelectronics on employment could either lead to disruptive conflict or the shaping of a new and better society. It is with regard to employment that the differences between the microelectronics evolutionists and revolutionists are most marked. The former argue that, as with all earlier technological developments, microelectronics, while causing a certain amount of transitional unemployment, will eventually expand product demand, open up new markets and create jobs. It has always been so in the past, why should it not be so now? The answer to any structural unemployment which may arise will be found through redeployment and in retraining schemes, although little indication is given as to where the new jobs are to be found. The revolutionists, on the contrary, suggest that the impact of microelectronics on all sectors of the economy simultaneously will greatly increase productivity and sweep away jobs across the board and hence traditional measures to deal with unemployment will have little or no effect. It is impossible at this stage to give an unambiguous answer as to which is more likely to prove true; it may be that future reality will be somewhere between the two extremes. It is clear, of course, that microelectronic technologies will create jobs in those industries which will manufacture novel electronic products and above all in the integrated circuit field itself. There are

also possibilities of growth in the manufacture of essentially new products such as electronic games, but these are not likely to be major
employers of labour and their markets will reach saturation. Again the
spread of computers will give rise to a high demand rate for programmers and those who produce the software; indeed here demand is
outstripping supply to the extent that it may slow down the spread of
use of the microcomputer. All in all, however, so many industries are
ripe for automation and so are many branches of the tertiary sector
that it seems probable that the net loss of jobs will be very
considerable.

It may well be indeed, that a sizeable proportion of the existing,
relatively high rates of unemployment is already due to displacements
caused by recent plant automation, but this is quite uncertain.

Over the last century, the outstanding example of major productivity
increase has been in the agricultural sector, in which technological
advances, combined with social and economic pressures, have greatly
reduced the work force engaged and have, at the same time, increased
production levels to the extent that now in none of the Western countries does agricultural employment account for more than 10 per cent
of the active population, while in some it is as low as 4 per cent.
Redundant agricultural workers were with little difficulty absorbed by
growing industries and they migrated off the land. Many believe that
this phenomenon of jobless growth can be repeated in industry and
that the surplus labour will be taken up by tertiary sector employment.
But historical analogies can be misleading, particularly when the circumstances are not exactly parallel—and that is the case here, because
the secondary and tertiary sectors are automating simultaneously. It
seems improbable, therefore, that redundancy from industry can be
mopped up by expansion of the service sector, at least in the traditional sense, although, as we shall argue later, it could be dealt with by
the creation of a new form of extended service of which we shall all be
part, at least for a proportion of our time.

Certainly the service sector has grown enormously in recent years,
mainly not as a result of the spending decisions of individuals, but of
the policies of the State supporting this expansion, notably in education, health care delivery, unemployment benefit schemes and the like,
in the capitalist countries through taxation. The great bureaucracies

which have resulted are, indeed, likely to be amongst the early candidates for automation and are, in any case, generally unpopular with the public, being regarded as cold, impersonal and inefficient. They are likely to be replaced by smaller, partially automated units of greater efficiency. Employment in transport and communications and large areas of the public service is expected to be affected by automation, unless productivity in these sectors is held at an artificially low level to maintain jobs, which will be difficult if the economy is to be efficient and competitive.

In the personal spending element of the service sector, the situation is quite different. As consumers have become more affluent, they have tended to spend more on vacations and especially on foreign travel to the general benefit of the tourist industry of many countries, on private medicine and education, on the impression that such services as provided by the State are of inferior quality. Employment in such areas is unlikely to be much influenced by improvements due to the microprocessor. Expenditure on personal service has not increased. As consumers have become more prosperous, they have not, as had been assumed, spent a higher proportion of their free incomes on "services" rather than on "goods". In fact, the high cost of personal service has pushed the trend in the opposite direction towards service-providing products; in practice they have purchased more and higher powered cars to the detriment of public transportation, domestic appliances to replace increasingly scarce and costly domestic help, and hi-fi, television and video-cassettes instead of going to a concert, theatre or cinema; i.e. labour-intensive services have been progressively replaced by capital-intensive mechanical and electrical equipment. This is one aspect of the automation of the service sector, seldom recognised as such.

The one hope of reducing the present high levels of unemployment is a massive and rapid expansion of world markets. A number of current trends seem to indicate that it will be difficult to return to the high rates of economic growth of recent decades; shortage or high cost of energy and imported raw materials, environmental constraints, an apparent reaching of saturation of demand in some countries for consumer goods and durables, and other factors, all seem to work in this direction. The only hope of returning, at least for a time, to high growth would seem to reside in increasing the prosperity of the less developed countries to an extent that they become major markets for manufactured and capital goods. In view of a lack of recognition of the positive advantages which such an event would have for the industrialised countries and their reluctance to move towards some sort of New International Economic Order, this seems a remote possibility. There would be, at the best, a slow development and it would necessitate a considerable redistribution of the international distribution of industry and labour and, as is indicated in a later chapter, the advent of the microprocessor seems to operate in the opposite sense.

There do seem to exist, therefore, strong reasons for concern that we may be entering a long period of considerable and probably endemic unemployment, a considerable proportion of which will result from automation made possible by microelectronics. One may well ask why, in face of this eventuality, nations should not slow down the rate of introduction of these new processes to allow the economy and employment to adjust to them gradually. The simple answer is that it is impossible. The very circumstances of economic malaise which we described means that each country will attempt to obtain competitive advantage through innovation in the search for extended markets. A longer term and more fundamental reason is that the microprocessor revolution offers eventually such enormous prospects of increasing riches, extending resources and of improving the quality of life universally, that it is impossible to reject it. In the early years there certainly will be a conflict between the need for innovation and high productivity and that of maintaining high levels of employment, but, as one trade union leader has said, "we know that the microprocessor will do away with jobs, but if we don't accept it, there will be jobs for nobody". The microelectronics era is with us for better or for worse. The main imperative seems to us to be that of looking ahead, analysing sector by sector and in the totality the risks and opportunities, and devising well in advance the social changes which will be necessary to achieve the benefits to the full. In the final paragraphs of this chapter we shall sketch one scenario for a possible approach; there may be other paths which could be useful to follow, but this one is presented on the basis of an optimistic view of the possibilities inherent in the microelectronic revolution for the transformation of society. The vision of a world free from poverty and from much of the burden of physical work is no new dream; it was in the minds of the fathers of the first Industrial Revolution two centuries ago. Then it was premature; today it is, at least technically, feasible, although its realisation is far from certain. Its achievement would demand understanding and foresight, wisdom and realism on the part of the political leaders, creative partnership based on common self-interest between government, management, the unions and science, and a high degree of awareness of what is at stake on the part of the general public. Decision will have to come within the next two decades if at all; either we shall seize upon these opportunities presented by the advancement of science and face up to the social adjustments inherent in their acceptance, or else our industrial societies will degenerate through inertia, social strife, and, above all, lack of courage.

At the heart of the matter is the question of employment. Concepts of employment, underemployment, unemployment and leisure are heavy with moral and historical values, some of them pejorative. In a situation in which large numbers of people are no longer required by industry, not as a result of cyclical fluctuations, but because society requires and technology makes possible very high levels of manpower productivity, these values have no longer significance and the words lose their meaning. This suggests that in the future the chief concern of the individual may be less employment as we know it today, but occupation in a larger sense, which will include certainly the time spent in contributing to the economic need of society, and for which he is paid, but also those activities, self-chosen, which provide personal fulfilment. Thus the occupation of the individual will have to be seen broadly as including a proportion of productive employment in the traditional sense, but presumably occupying a much smaller part of his life (later entry after more education, shorter hours, earlier retirement, periods free for further education or training), together with one or more subsidiary occupations of a craft, artistic, educational or other nature, in addition to leisure in the traditional sense. The "economically productive" element of the individual's occupation would include periods free for re-education and for the deepening of skills in the subsidiary occupations, the acquisition of techniques, practice and experience under a teacher. The secondary occupations would be organised, encouraged and made freely available on a local basis, with tools and courses of instruction provided indirectly by the State, but, of course, on a completely voluntary basis, it being recognised that it is in the interest of society as a whole that a very wide range of interesting and constructive alternatives is available to meet the diversity of human needs and to prevent the alienation which could result from excessive isolation in front of the television screen and the desk computer.

Education would have to be modified considerably to provide a basic capacity for learning, rather than for the acquisition of information which would be so freely available throughout life in any case through the new media. In the later stages of formal education, a main objective would be to lay the basis of understanding and provide the intellectual skills required to build, maintain and further evolve the automated industries which would provide the resources supporting the whole system. Unemployment in the present sense, with is pejorative and demoralising image, could cease to exist.

Adoption of such social policies would entail a basic rethinking of the work ethic on which the material prosperity of the presently industrialised countries has been built. But this would have to take place in any case if the microprocessor society does give rise to massive and endemic unemployment. There would also be quite fundamental repercussions on the political systems and their ideologies.

The scenario is not so radical or improbable as it may seem at first sight. If the development of microelectronics in speeding up the automation of industry and of office work does indeed generate great and intractable problems of unemployment and if organised labour recognises that it cannot, as a matter of longer-term self-interest, reject the process of automation, negotiations will follow, in which governments as well as the unions and the employers will take part, towards an equitable distribution of employment, through shorter working hours, earlier retirement and other means so as to prevent the existence of an unacceptably large pool of permanent unemployment. In addition, with shorter periods of work for all and with the need to counteract the demoralisation of the permanently jobless, it would seem necessary to make easily available social occupations on a voluntary basis to render the increased extent of free time of people creative and satisfying. Such obvious measures could well lead the way into a new type of "occupation" society such as the one sketched above, but in addition much thought will have to be given to the construction of broad policies of income distribution.

As a final and controversial coda to this Introduction, let it be added that in the social situation described, as always, the solution of one set of problems leads to the arising of others of a different nature. Even Utopia will not be without its difficulties and conflicts, given the persistence of human nature. Behind the not impossible vision of a new society, which has been described, rises the shadowy question as to whether or not the human, in common with the other species of creation, can continue to flourish and to resist degeneration and atrophy without the struggle for existence and the compulsion of work.

*   *   *

It has been felt useful to begin this book by looking at some of the longer term perspectives of the development of microelectronics as a framework within which the difficulties and benefits, more immediately apparent, can be discussed. Many of the matters touched on are dealt with in greater detail in the chapters which follow, where the arguments are more fully elaborated.

The first part consists of a chapter by Ranald Ide on "The Technology" of microelectronics, followed by that of Ray Curnow and Susan Curran entitled "The Technology Applied". After these descriptions follows a part on socioeconomic aspects, consisting of three chapters, that of Bruno Lamborghini: "The Impact on the Enterprise"; that by John Evans: "The Worker and the Workplace"; and finally the chapter by Günter Friedrichs: "Microelectronics and Macroeconomics". Then follows a chapter by Juan Rada entitled "A Third World Perspective". Many readers may feel that the book too closely reflects the preoccupations of the industrialised countries of the First World. This is to some extent inevitable, since the development of microelectronics has taken place essentially in these countries and the impact of its applications will be first met there. The authors of the book have, throughout their discussions, been deeply concerned with the probable impact of this development on the less developed nations and great importance is given to a full appreciation of this aspect.

The chapter by Frank Barnaby on "Microelectronics in War" seems to the authors to be essential in understanding the full significance of microelectronic applications. It underscores the negative possibilities of the use of this, as of other scientific developments, and is a strong argument for the need for a total approach to policy-making in this new field.

The next chapter, dealing more specifically with some aspects of microelectronics in shaping "the information society", is by Klaus Lenk entitled "Information Technology and Society".

Finally we return to the global and political perspective in a chapter by Alexander King entitled "Microelectronics and World Inter-dependence" which again evokes the North-South issues, before closing with a postscript by Adam Schaff who reflects on the abolition of work under the title "Occupation versus Work", considering more profoundly some of the thoughts expressed in this Introduction.

2

The Technology

THOMAS RANALD IDE

Information as a Characteristic of Modern Society

The relationship between information, computers and microelectronics is often taken for granted, particularly by those who work closely with the technological developments related to communications. Information, of course, in its broadest sense relates to any and all facts that are communicated, learned or stored. Information theory is more precise and concerns the process or storage of data. Information technology, for its part, refers to the means by which information is handled. In France, "Informatique" originally referred to the science of information; however, because of its close links to computers, it is now understood to mean the science by which computers process and store information.

The requirement for more information, particularly information of a highly complex nature, together with the high costs of accessing and processing it in non-electronic form, has created a demand that has effectively complemented the technological push evolving from accelerating developments in computation. These are a direct result of the introduction of microelectronics into these areas.

In the industrialised world we have moved increasingly towards an information society based on a microelectronic technology which is both capital- and labour-saving in its applications. It is this technology and those others directly related to it that this chapter proposes to analyse.

The Art of Computation

The means of computing have been a major preoccupation of human beings since the dawn of civilisation. The Pyramids and Stonehenge are surviving monuments to the talents of early mathematicians. By and large the systems they used were either *anologue* or *digital*. The first, as its name implies, is based on analogies. The thermometer is a prime example in which increases and decreases in temperature are measured by corresponding changes on the length of a column of mercury. The slide rule is also an analogue device, with the products and quotients of numbers related to analogous lengths on the rule. Digital devices, on the other hand, related to counting numbers. The most common system, the decimal, is based on ten digits and probably evolved from the custom of using the fingers for calculations. But just as a system based on ten can provide for arithmetic functions, so can a system based on any other number.

The binary system requires only two symbols, that is "1" and "0", and is fundamental to the operation of the modern computer. These symbols are not limited to numbers. If used for "on" and "off", they represent the two positions of an electric switch. They also can stand for other alternatives such as "yes" and "no", "and" and "or", etc. When George Boole developed an algebra whereby the relationship between a large number of elements could be dealt with by repeated applications of the relationship between a series of two elements, the ground was laid for today's advances in the field of modern microelectronics.

Space does not permit a detailed look at the history of the development of computers and communications devices. However, even a brief outline reveals something of both the nature of the modern computer and the rate at which new developments have been and are accelerating. It also hopefully will serve to demystify processes which, while not simple, are the products of the curiosity and ingenuity of human beings.

Historical Developments

Early developments

The first digital mechanism, the Abacus (Plate 1), dates back to at least 3000 BC, and is still employed effectively in many parts of the world today. No further major development occurred until 1642 when Blaise Pascal in France, at the age of 19, built the first simple digital computer capable of addition and subtraction. In 1672 he was followed by Wilhelm von Leibniz in Germany who used stepped wheels to construct a machine which not only could add and subtract but also multiply, divide and extract square roots.

Over 150 years later, in 1835, the Cambridge mathematician Charles Babbage designed a machine which, although never built, earned for him almost universal recognition as the father of the modern computer. It contained input and output devices which made use of punched cards similar to Jacquard's master loom (Plate 2) which employed them to create a predetermined pattern. In addition, Babbage provided for a store or memory and a mill or processor. The store and mill were also to be governed by cards on which instructions were coded numerically and stored until required for operating the processor.

Analogue and digital mechanical computers

In 1915 Leonardo Torres in Spain combined electric-mechanical calculating techniques with principles of programming. He demonstrated the first machine capable of making decisions and illustrated its versatility by applying it to the solution of simple chess problems. Sixteen years later in the United States Vannevar Bush designed an analogue computer which he called a differential analyser. This was the first computer with the general capability to solve equations. Two years later in the United Kingdom Douglas Hartree combined with Arthur Porter to build an analogue computer using about £20 worth of Meccano parts. Hartree later became the first scientist to use a computer to solve problems in atomic theory.

About the same time (1936) in Germany Konrad Zuse pioneered

some of the basic ideas of automatic computing, including the use of the binary system and the floating decimal point. By 1941 he had completed his relay calculators Z2 and Z3 and had developed an algorithmic language PK, the forerunner of PL/1 and ALGOL 68.

In 1937 Claude Shannon and George Stibitz, working separately in the United States, were able to devise electrical switching circuits which worked according to Boole's principles. As a result, the technology of computation was significantly advanced.

Seven years later the Harvard Mark 1 Calculator was demonstrated. This, the first all-purpose digital computer, was jointly developed by Howard Aiken, Clair Lake, Francis Hamilton and Benjamin Durfee working at the Harvard Computation Laboratory with IBM support.

The vacuum tube era

The use of vacuum tubes (valves), invented by Fessenden, DeForest and others in the early 1900s to control electric currents, began to find applications in computer development in the early 1940s. In 1945 John von Neuman, born in Hungary but working in the United States, developed the concept of the stored programme, whereby instructions for the computer could be retained internally in numerical form. As a result, logical choices could be made inside the machine and the instructions modified by the computer during the process. This was a critical step which owed much to Babbage's work of over 100 years earlier and led, in 1949, to EDVAC (Electronic Discrete Variable Automatic Computer) at Princeton, ENIAC (Electronic Numerical Integrator and Calculator) at Pennsylvania and EDSAC (Electronic Delay Storage Automatic Calculator) at Cambridge.

ENIAC was the first all-electronic computer. Designed by John Mauchly and Presper Eckert, it was completed in 1946 and reputedly cost close to $10 million, contained 18,000 vacuum tubes and weighed approximately 30 tons (Plate 3).

New and increasingly sophisticated computers followed. IBM's "Poppa" in 1948 included conditional transfer of control, and EDSAC in 1949 was the first to achieve a high-speed memory using binary numbers. SSEC, EDVAC, ILLIAC, MANIAC, "Whirlwind", MADM and UNIVAC brought to an end the development of the extremely large, expensive machines in the early to mid-1950s.

The transistor and integrated circuits

Up to this point progress had been comparatively slow. Computers were not only costly but also required large amounts of space and equipment to design and build. However, with the discovery of a small, solid-state device this was to change dramatically.

In 1948 John Bardeen, Walter Brattain and William Shockely at Bell Laboratories in the United States developed the transistor which, because of its small size, reliability and low power requirements, prepared the way for integrated circuits and the advent of the age of microelectronics.

In 1951, at Western Electric in the United States, scientists developed the first solid-state amplifier—the Point Contract Transistor.

In 1958 Fairchild in the United States developed a planar transistor made using silicon dioxide as an insulator.

In 1959 both Texas Instruments and Fairchild developed semiconductor packages with two or more transistor devices within a silicon substrate or base. As a result, costs were significantly reduced, and the way was opened for an ever-increasing number of components to be contained on a piece of silicon. In 1964 Gordon Moore, when he was research director at Fairchild, estimated that density[1] should double each year. Some sixteen years later the prediction has roughly held, although some modification in the rate of growth is beginning to appear.

In 1971 Intel in the United States, now a major manufacturer of integrated circuits, developed the first microprocessor, that is a central processing unit (CPU), where the logic and arithmetic functions can be performed on a single silicon chip of less than 0.5 centimetre per side. In 1975 the same company managed to assemble a complete computer on a single printed circuit board. One year later it announced the production of an eight-bit computer consisting of 20,000 transistors on a single silicon chip.

---

[1] Density refers to the number of transistors per unit area of silicon and in the jargon of computer scientists is often called "real estate".

Microelectronic Technology

Context

Much of the discussion in this book relates to the impact that computers and communciation media are having and may be expected to have upon the various groupings into which mankind has become divided. Computers, satellites, optical fibres and their applications in the factories, the offices and in the home are the visible elements which we will utilise for our well-being or otherwise. The engine which is driving these developments is small, extremely inexpensive when mass produced, modest in its energy needs and very powerful. It is the microprocessor. It and the technology associated with it is commonly known as microelectronics. The principles on which it is based, while not necessarily easy, are far from difficult to understand. They include, as well as elementary electronic theory, the binary system and Boolean algebra.

Microelectronics

Microelectronics is a subset of electronics, an applied science which deals with the motion of very small subatomic negatively charged particles called electrons and covers their behaviour in gases, vacuums, conductors and semiconductors. Electrons move in an electrical field, thus forming a current. The circuits they follow contain components which are either active (transistors) or passive (resistors, capacitors, inductors).

The unusually rapid growth of microelectronics would have been impossible without the development of the transistor. By replacing the large and inefficient vacuum tube, the way was opened to the miniaturisation of the computer. It effectively complemented the requirement for complex stored programmes that were to be a feature of the space age where size, reliability and minimal energy needs are of critical importance.

Integration of numbers of transistors and other passive components such as resistors was always a possibility, but had to await the development of methods for refining silicon to the necessary degree of purity and techniques which permitted layers of "impurities" to be placed within the silicon crystal. It is these impurities which form the components. The design of the circuits are transferred to successive layers using stencil-like masks, which are first drawn to a relatively large scale and then reduced many times by photographic processes. Advances in photolithography[2] and other techniques of miniaturisation made realisation practical.

Once these techniques had been developed, it became possible to imprint thin slices, or wafers, of crystal, usually silicon, with several individual, identical integrated circuits. These wafers, several centimetres in diameter, are then cut into small, rectangular pieces of about 0.5 cm a side called chips each containing an integrated circuit. It is the integrated circuit that is the basic unit of microelectronic technology.

A transistor may be compared to a light switch which can be turned on and off. If used for digital applications, it is like a simple electrical switch which is either "on" or "off", or in logical terminology the "1" state or the "0" state.

An integrated circuit can be compared to an electrical control panel on which all the switches and other circuit components, such as resistors and capacitors, have been mounted in a compact and well-organised manner. The panel is then miniaturised from desk top size down to an area roughly a fraction of the size of a postage stamp, so that each switch occupies only a few millionths of a metre on a side, and the connections between them are made at two or more levels in a grid buried in an insulating layer covering the switches. The panel still requires ground and power connections as well as input and output lines.

A transistor consists of three elements—a base, a collector and an emitter. A weak current flowing into the base and out of the emitter controls a much stronger current between the collector and the emitter (Fig. 1), thus the current is amplified. The layers are created by im-

---

[2] Photolithography uses ultraviolet light which shines through a mask containing a circuit pattern to expose a layer of polymer which coats the silicon wafer. This can then be selectively etched using acids, allowing impurities or metals to be applied, where required, on the wafer.

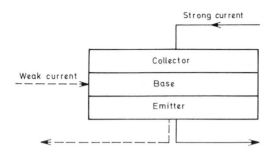

Fig. 1. Schematic diagram of the transistor.

planting impurities in the pure silicon. The type of impurity in each level is different. For example, phosphorus supplies extra electrons and hence is a negative (n) layer, whereas boron absorbs electrons and hence is positive (p). Thus the conducting properties are controlled and determine the direction in which the current flows. This provides for the switching capability.

The majority of transistors used today consist of a pair of back-to-back diodes, that is devices which contain positive and negative electrodes. The electrical switch action is achieved by applying an electrical impulse to the interface between the diodes. In one of the two most common types of transistor, the bipolar (Fig. 2), a resistance is

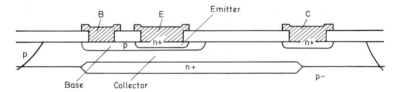

Fig. 2. The bipolar transistor.

set up in the base and electrons entering it from the emitter are carried across to the reverse biased base collectors. Such a transistor can act as a high gain amplifier or as an "on-off" switch.

The second principal transistor (Fig. 3) is known as the Metal-Oxide-Silicon Field Effect Transistor (MOSFET). In this instance a conductive field plate is placed on top of the layer over the gate area.

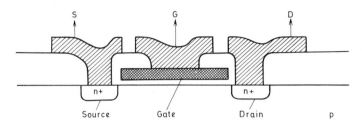

Fig. 3. The MOSFET transistor. Source: Bell Northern Research, Canada.

When an electrical charge is applied, the current is allowed to flow. It is a very effective amplifier, and as a switch it can be easily used to provide for the logic elements of Boolean algebra. It is energy efficient, since current only flows when the switch is "on". This is the basis of the Complementary-Metal-Oxide-Silicon (CMOS) which at present appears will be a major feature of development in the next decade.

Transistors, the fundamental component within integrated circuits, can thus amplify signals or switch currents on or off.

The number of components that can be placed on a single chip has been increasing at an exponential rate. In 1965 the number of transistors that could be accommodated was about 10; by 1980 the 10,000 transistor chip was common. Other components, that is resistors, capacitors and diodes, are also fashioned in the chip and these, with transistors, now total close to 100,000. Component density is then a convenient measure of the power and capacity of a chip.

Depending on the number of "logic gates" and hence the number of functions that can be performed, integrated circuits are further classified as small-scale integration (SSI), about 10 components, medium-scale (MSI), from 64 to 1024 components, large-scale (LSI), from 1024 to 262,144 components, and very-large-scale (VLSI), beyond 262,144 components.

The binary system, Boolean algebra, "bits"
"bytes", "baud" and "memory"

The number of digits needed to represent a number is the number of units in the base. Two digits, 0 and 1, are all that is required in the binary system; six, that is 0, 1, 2, 3, 4 and 5, in the sexagesimal system; and ten in the decimal system. For example, the decimal numbers 0, 1, 2, 3, 4, 5, 6, 7, 8 and 9 are represented by the binary as 0, 1, 10, 11, 100, 101, 110, 111, 1000 and 1001 respectively. Through this counting system any number can be shown as a string of "0s" and "1s", and all arithmetic functions can be reduced to operations on "0s" and "1s".

Using the principles developed by George Boole, binary numbers can also be used to express propositions in symbolic logic. Analysis of a problem is carried out by a series of declarative statements which are either "true" or "false" along the lines of the popular parlour game "Twenty Questions". In electronic circuitry two propositions that are either true or false, but not both, are each associated with circuit elements called "logic gates", which will close if the proposition is true or open if false. Thus it is possible to perform a variety of functions related to the operations *and, or* or *not. Not,* as the word implies, simply negates the proposition. If, for example, given two statements, P and Q, "P *and* Q" is only true if both P and Q are true simultaneously, whereas "P *or* Q" is true if either P or Q is true. Since some operations or functions are complicated, the circuitry required is complex, in some cases using thousands of logic gates.

The basic units of information are called "bits", which are analogous to the "1" and "0" of binary arithmetic and the "true" and "false" of Boole's algebra. In addition, bits can be used to represent letters, characters, words and sounds. It takes eight bits, a "byte", to represent a character, as eight bits can represent $2^8 = 256$ different possibilities, enough for the Latin alphabet, the digits 0 - 9 and the common punctuation symbols. All words in the conventional sense can be broken down into forty basic speech sounds called "phonemes". Each phoneme is a set of frequencies with a certain sequence and is expressed electronically as a set of "0s" and "1s".

The memory capacity of a computer is usually calculated in bytes. A "baud" is used to describe the speed of information in a telecommunications channel, and is normally equivalent to "bits per second". Byte and baud are the jargon of the computer technologists. Bit, however, as the basic unit of information has achieved virtually universal acceptance. Unless otherwise specified, it will be used throughout this section.

Anatomy of the computer

While computers have become increasingly efficient since Babbage first described his "difference engine", they still require the same basic elements included in his design. For example: *a programme* to instruct them regarding the operations to be performed; *input devices* such as cards, tapes, keyboards, microphones and electronic readers through which the programme and other instructions can be fed to the computer; *storage* and *memory* capacity which receives and stores the instructions and such other information as the user wishes; a *central processing unit* to carry out the instructions provided from the storage; and an *output and display device* such as a cathode ray tube (CRT), printer or speaker. There are other elements, but in each case they relate to the functions described above.

Memories — storage

Advances in microelectronics have had a dramatic impact on the memory or storage capacity of the computer of the 1980s. Indeed it is likely that by the end of the decade the contents of thousands of books may be stored on a disk using a laser-based optical system.

The means by which information is stored and retrieved is commonly characterised as primary or secondary storage. Primary systems are solid state semiconductor devices, i.e. memory chips. While they are comparatively limited in their storage capacity, the information they contain can be accessed in a fraction of a millionths of a second. They are also small and hence are ideal for employment within the computer. Secondary systems consist of external magnetic disks capable of storing large amounts of information. These are of two types—rigid or flexible (floppy). The rigid are more expensive, but have a significantly larger capacity—up to a million bits per square inch compared to three to four hundred thousand for the flexible. They are also faster, with transfer rates of up to 10 million bits per second compared to two or three hundred thousand for the floppies. Magnetic tapes are inexpensive, but their capacity is limited to under a hundred thousand bits per cassette, and the access time is slow, around 4000 bits per second.

While magnetic systems have dominated the market to date, the development of optical systems, whereby information is stored and retrieved on and from disks by lasers, promise capacities which will exceed those of their predecessors by a thousand or more times. In addition, the analogue signals generated by sound and pictures can also be converted to digital data and thus be stored and retrieved optically. Advantages lie not only in the increased capacity provided, but also in the virtually error-free nature of the optical process.

Memories are also classified according to the nature of the accessibility of the information. A read-only memory (ROM) is preprogrammed with the information it contains readable only and not able to be altered. It is ideal for certain applications, since it cannot be erased. The information it contains is integral to the operation of the computer. Random access memory (RAM) is extremely flexible, since the information can be altered or "rewritten". Unfortunately the information is erased automatically with a loss of power. Programmable read-only memory (PROM) is another form of memory chip and unlike the ROM it can be programmed by the user. A further refinement has been introduced with erasable programmable read-only memories (EPROM). Here the user retains the ability to modify or change the information on the chip at will.

Although invented at Bell Laboratories as long ago as 1966, recent improvements have created new interest in bubble memories. These are semiconductors which have been exposed to a magnetic field which generates hundreds of thousands of tiny circles (bubbles), the presence of which is charged to represent a "1" and the absence to represent a "0". Essentially they are a semiconductor equivalent to a secondary system, but which promises to be cheaper than any mechanical device. One laboratory experiment achieved a storage of one billion bits on a 15×18 cm board with a transfer rate of 100,000 bits per second. Devices marketed today store and move a million bubbles within a film on a chip with an area of less than 1 square centimetre. Future developments using magnetic garnet film could multiply this capacity four or five times.

Competing with magnetic bubbles and charge-couple devices (CCD) which store and process binary code information in the form of packets of electrical charges, CCDs are slower, simpler and cheaper than the conventional solid state semiconductor standard memory chip. As with magnetic bubble memories, they have no moving parts and hence are more reliable and less likely to be damaged than disks or tapes.

The efficiency of chips is often measured by their random access memory (RAM) capacity. The number of bits of information stored is roughly equivalent to twice the number of components. That is, a RAM chip capable of storing 65,536 bits of information requires approximately 128,000 components. Because of the size of the numbers involved and because the integrated circuits must work with binary numbers, the "kilobit" (1 K) has become the basic unit where 1 K equals 2^{10} or 1024 bits. By 1980 a number of companies were producing 64 K RAM chips. A 256 K RAM chip is expected to be tested in the near future. This will imply a similar increase in the density of the components and signal the advent of the very-large-scale integration (VLSI) era.

Microprocessors and the microcomputer

The arithmetic logic and control operations are performed in the central processing unit (CPU) of a computer. In the age of microelectronics the CPU has become the microprocessor, with its components contained in a single chip. Its parts must communicate with one another, and this is accomplished through a series of conductors called a "bus", that is a number of parallel conducting paths which not only form the internal communication system but also extend through connecting pins to a set of parallel conductors outside the chip. In addition, the microprocessor includes units to interpret instructions from the stored programme to supply the control memory with the information necessary to retrieve instructions and to send out data as required.

In contrast to the microprocessor, the microcomputer is virtually a complete system within itself. In addition to the functions performed by the microprocessor, the microcomputer contains control chips which ensure that the electronic signals flow smoothly and in their designated order through the complex circuits. Other chips provide the primary memory requirements and the input-output functions. Microcomputers are usually contained on circuit boards with dimensions of about 20 cm × 30 cm; however, a complete microcomputer on a chip became a reality when Intel in 1976 managed to crowd 20,000 transistors on a single chip with an eight bit capacity.

Microprocessors and microcomputers are rated by the number of bits in the information processed. For example, an eight bit microprocessor is limited by the quantity of information it can process in a given instruction cycle, that is numbers up to 256 or equivalent data. On the other hand, a sixteen bit microprocessor has much more power and can handle numbers up to approximately 65,000 in the same unit of time[3], but it also requires about ten times the number of components per chip, close to 100,000 compared to about 10,000. The relationship between component density, random access memory (RAM) and the word length of microcomputers is shown in Fig. 4.

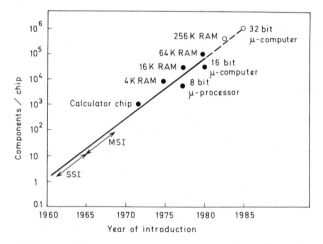

Fig. 4. Evolution of the microcomputer. Source: Morgan Stanley Electronics Letter.

[3] The numbers relate to powers of 2; 8-, 16- and 32-bit microprocessors = 2^8 (256), 2^{16} (65,536) and 2^{32} (4.295 × 10^9) respectively.

Microprocessor are also rated by the speed with which they can process information. The units used are the fractions of a second taken to process the simplest operation. Obviously the shorter the distance the signal must travel, the shorter will be the time required. The time taken to access information in the memory is a major factor. Thus an increase in the density of components on a given chip tends to increase the speed of operations. Secondary memory devices such as tapes and disks are useful for storing programmes and data, since their capacity is large even though their access times are slow.

Large, expensive computers, sometimes described as "mainframes", have also been affected by miniaturisation, as increasingly they are made up of groups of microprocessors. As a result, their capacity has been enhanced, energy requirements lowered and outputs dispersed with terminals close to the user.

The distinction between mini- and microcomputers is becoming blurred as more and more peripherals or modular additions are being designed for microcomputers, and at the same time better means are being developed for interfacing the two categories of computers. It is probably sufficient to say that the minicomputer is a small but high performing machine with mass storage capacity and special features costing $25,000 or more in 1980 terms.

The microcomputer, in comparison, has less built-in capacity and hence is cheaper, with some models selling for as little as $500 (1980). The basic unit consists of a packaged printed circuit board containing a microprocessor, some memory chips and input-output connectors. The externals provided are usually a keyboard for input, a video display for output and a tape recorder for storage. Interest in the applications has grown to such an extent that a number of additional peripherals are obtainable. These include printers, facsimile machines, modems,[4] voice and music digital synthesisers, plus disk drives, both floppy and hard, for additional storage and software. Depending on the extent of the system, costs may go above $15,000 (1980). By connecting microprocessors in such a way that they are able to com-

---

[4] A modem, the acronym for modulator-demodulator, encodes and decodes information as audible tones that can be transmitted directly over a telephone.

municate with one another and to other data bases, the applications of the microcomputer can be extended almost infinitely.

Software and languages

Software is a frequently misunderstood term. Since computers and other equipment are often referred to as hardware, it seems natural to assume that software means content. And in a limited sense it does. But to the computer scientist it is the operative programme, the long lists of commands or instructions that tell the computer what to do, which are supplied either by the manufacturer or someone who specialises in programme design. The instructions must be precise and explained with the most careful attention to detail if the computer is to function correctly.

Firmware is another term that has crept into the jargon. In this case the reference is to fixed programmes built into the computer at the time of manufacture. As software becomes more and more expensive, the need for and value of firmware increases.

Languages, on the other hand, prescribe the way the user communicates with the computer. Machine languages are based on the binary system which the computer understands. But such languages are not suitable for the user because they are incomprehensible to humans. Software also is often written in "assembly" languages which are relatively close to the machine language and are translated by a programme known as an "assembler" into machine language. The typical user, however, is much more at ease with languages that are close to his or her own. A number of these have been developed and are commonly known as "higher level languages". The translator in this case is a programme known as a "compiler". Higher level languages are oriented toward the problem they are designed to solve. They include FORTRAN or formula translation, COBOL or common business oriented languge, BASIC or beginners all-purpose symbolic instruction code and PEARL, a high level, real time, process control programming language which is coming into increasing use in the Federal Republic of Germany.

Software then refers to sets of instructions that can be designed or acquired. They can range from professional packages such as medical and surveying to business packages such as inventory and payroll. Partly because the development of the hardware has been so rapid and partly because the creation of software is so highly labour-intensive, there is a software shortage worldwide in extent. This shortage is reflected in an analysis of data processing costs as shown in Fig. 5.

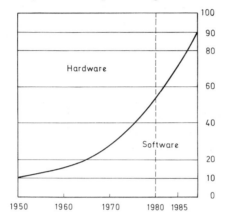

Fig. 5. Data processing costs (e.g. in 1950 hardware represented 90 per cent of costs while in 1985 it represents some 20 per cent). Source: Boehm, Datamation.

The development of firmware was expected to assist in overcoming the problem, and it has to a degree, but not to the hoped for extent. Another approach to the problem was to encourage the use of high level languages for software development, but even this has not enabled the supply to meet the demand.

Languages are the medium by which the user communicates to the machine. Low level languages are those used by the computer, whereas higher level languages are as similar as possible to that of the user.

One of the problems that has grown almost as rapidly as computer technology itself is the lack of compatibility among computers. The increasing number of different languages is an integral part of that problem. Without an appropriate piece of software or interface programme it is becoming increasingly difficult for users to take advantage of the substantial amount of useful information which is being stored in the data bases associated with the different computer systems. Even where the software interface exists, it is sometimes difficult to ensure full compatibility. There are advantages associated with plug-in compatible machines and some manufacturers are moving, albeit slowly, in this direction.

Microelectronics and the cost factor

A 1979 OECD Report referred to the introduction of microelectronics as a qualitative change and stated that electronics and its applications will be central to the reorganisation that will necessarily take place within the production structures of advanced industrial societies. This opinion reflects not only the increases in productivity which result from the electronic rather than the machine mode but also the rapidly declining costs associated with microelectronic components. Computers are not only becoming smaller and vastly more powerful, they are also becoming cheaper.

As we have seen, the power of computers has increased during the past fifteen years by almost 10,000 times, while the price of each unit of performance has decreased 100,000 times. The computers of 1963 required several tens of thousands of man-made connectors, all of which were capable of failure. This compares to the equivalent of less than ten elements in today's large-scale integrated circuit boards. That this trend is continuing and is expected to extend into the future can be seen from the following.

Figure 6 shows the growth rate in the number of electronic components on a microcircuit chip, the number of bits of information store in random access memory (RAM) chips, and the number of bits in the word length of microcomputers generally. It is interesting to note that the number of components per circuit has increased from 256 in the mid-1960s to a thousand times that many ten years later.

In Fig. 7 the decline in the cost per bit of computer memory is displayed for successive generations of RAM memory circuits, while in Fig. 8 the annual utilisation of electronic functions is shown. By 1985 it is expected to increase by some 2000 times. The reduction in costs for logic gates as they relate to the use of vacuum tubes, discrete transistors, small-scale integrated circuits and large-scale integrated

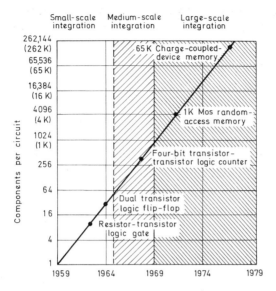

Fig. 6. The evolution of integrated circuits. Source: Noyce, R.N., *Scientific American,* September 1977.

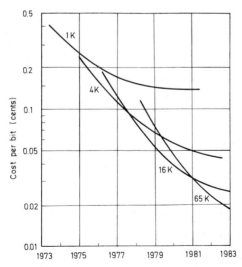

Fig. 7. Cost per bit of computer memory for successive generations of RAM. Source: Noyce, R.N., *Scientific American,* September 1977.

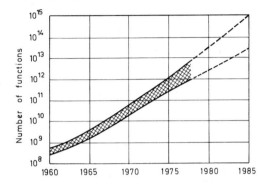

Fig. 8. Annual utilisation of electronic functions (transistors, logic gates, memory bits). Source: Noyce, R.N. *Scientific American,* September 1977.

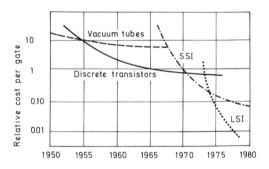

Fig. 9. Reduction in costs of logic gates. Source: Mayo, J.S., *Scientific American,* September 1977.

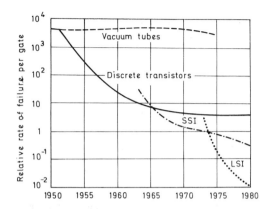

Fig. 10. Gains in reliability of electronic components. Source: Mayo, J.S., *Scientific American,* September 1977.

circuits is shown in Fig. 9, while in Fig. 10 the gains in reliability for the same components are displayed.

It is because computers are cost effective and are able to accomplish tasks which were beyond the ability of society only a few years ago, that their growth rate has been so phenomenal. This rate is unlikely to change. On the contrary, as they become smaller and cheaper they will become more ubiquitous and will intrude more and more on the human scene. As they become smarter and faster, they could well take over a greater and greater share of the decisions that individuals have been accustomed to make for themselves.

Microelectronics and Communications

The digital equation

It is not only in the area of computer technology that costs have decreased and capacity increased so rapidly. Similar trends characterised the communications field. When economic means were found to transmit large quantities of alphanumeric data (letters and numbers) in a digital format, the two technologies began to complement one another.

Information in digital form also enjoys significant advantages in cost of operation and reliability of performance. Since the representation of all information, whether voice, video or data, appears as a series of bits, the way in which the information is used, transmitted or switched is independent of the original source or format. Furthermore, signals transmitted in digital form are not nearly as likely to suffer a quality loss or degradation.

Digital techniques also make time-sharing on telephone cables feasible. Known as time-assignment speech interpolation (TASI), the concept makes it possible to use the natural breaks within speech for the insertion of data from other sources. So effective is this device that when speech is retranslated into its analogue form at the receiving end, no interruption or interference is perceived. The integration of voice and data into a single system has permitted interface with different networks. Efficiencies have been achieved and costs have fallen accordingly.

Packet switched systems also owe their development to the introduction of the digital mode. The control computer is able to identify routes where traffic is relatively low and direct packets of data along them even though the distance to the designated reception point may be greater. As with TASI, the information is reconstituted at the end of the transmission. The ability to ensure that the lines are better utilised means cheaper and faster service.

The major advantage of the digital mode lies in its compatibility with integrated circuits and microprocessors. Their small size and weight are particularly important to communications devices. Without the size and weight advantages associated with microelectronics, the communications satellites of the space age could not have been launched.

These applications to communications technology have naturally resulted in a multiplying effect on the microelectronics industry. Developments in one area stimulate developments in the other.

Voice synthesis

Reference was made earlier to phonemes and their use in translating audio into digital data. Some years ago one manufacturer created and marketed a game designed to teach small children to spell. The vocabulary was limited, but the device demonstrated effectively the capacity of a computer to speak. Improvements were rapid, and today a computer, the Kurzweil Reading Machine (KRM), is capable of reading books aloud. Others can articulate instructions to a forgetful car driver and will eventually be able to take dictation and produce a typed draft within a few seconds.

Regarding the problems of voice recognition and reproduction, recognition is by far the most difficult; reproduction is relatively simple. Two approaches have been taken. The first is an attempt to recreate what goes on in the vocal tract. This includes not only the characteristics of the vocal chords but also accompanying noise such as aspirations and radiations caused by the lips. In this method the principle units are "formants" which represent sounds in terms of their resonant frequencies and relative amplitudes.

The second system is to create a model of the human voice and develop an appropriate technique to describe it. Known as linear predictive coding (LPC), approximately twelve samples are taken, and because speech follows a reasonably predictable pattern, the model is created. The units used in this case are phonemes, each of which represents a typical sound, and using a digital/analogue converter the necessary sounds are created.

For reproduction, one kilobit microprocessor chips are capable of producing good quality speech with virtually unlimited vocabulary. The constraints lie in the time required to process the data. In the mid-1970s minutes were required to process 1 second of speech. By the end of the decade this had been reduced to 15 seconds. The words themselves are stored in standard memory chips, and their number is limited only by the capacity and the quantity of the chips.

Speech recognition had been achieved by the end of the 1970s for limited vocabularies of a few hundred words and a limited number of voices. However, developments have proceeded rapidly in the creation of the necessary algorithms, that is the methods describing the sequence of steps the microprocessor must follow to identify the multitude of variations that characterise oral communications.

The major deterrent has been the amount of computer capacity required to meet the demands of the algorithms. The advent of the thirty-two-bit chip and its likely commercial availability in 1983 or 1984 terially assist to overcome the obstacle relating to the size plexity of the message.

ne mid-1980s the increased computer capacity should make ble instantaneous oral reservations and confirmations with nes, railroads and hotels, etc. Further, by the end of the century an individual could well be able to dictate sophisticated data directly to an electronic typewriter and receive copy correctly punctuated and free of spelling and typing errors.

Satellites

Satellites are an example where developments in communications have stimulated developments in microelectronics. In 1965 a single satellite carried 240 telephone circuits; today (1980) the modern version carries 12,000. In 1965 the cost per circuit per year was $22,000: today it is $800. By 1985 the next generation should carry 100,000 circuits at $30 per year. A similar development is occurring in associated ground stations. In 1965 one that could both transmit and receive was priced at approximately $10,000,000. Today the largest costs less than $300,000, and small, receive-only parabolic dishes are presently selling in Japan for about $200.[5]

The falling costs of ground stations have important implications for the dissemination of information and entertainment. Satellites are virtually cost-insensitive as to distance, hence individuals in remote and rural areas have opportunities to access transmitted material as cheaply as the urban dweller in the industrialised nations. Microelectronic developments have made a major contribution because of their small size, significant capacities and modest energy requirements. In general, the higher the power of the satellite and the higher the frequencies used, the lower the relative cost and the smaller the size of the ground stations. Such satellites are commonly referred to as direct

---

[5] The above figures are given for domestic satellites such as those operated by Indonesia, Canada and the United States. Intelsat, a respected international consortium, provides service designed for global use, hence its satellites are located over ocean centres. Thus the signal, as delivered, is much weaker and requires ground stations to be larger and more costly.

broadcast satellites (DBS), and larger and larger numbers of individuals are acquiring their own receiving dishes. Their use was pioneered in Japan to meet the needs of the inhabitants of its northern areas. Canada has also been active, spurred by the need to provide service to the small population in the huge northern sections of the country.

A number of countries have launched space vehicles of one kind or another since the USSR first put Sputnik into orbit in 1957. By 1980 they numbered in the hundreds, including those in geostationary orbit[6] which provide the communications links to associated earth stations. Other applications apart from communications include atmospheric probes, remote sensing to assist in the monitoring and management of resources, search and rescue information assistance, meteorological data acquisition and distribution and surveillance satellites of both a military and non-military nature.

There are a number of technological developments related to satellites which will provide new opportunities for access to information. Some of these, such as transborder data flow, raise questions relating to economic integrity, national sovereignty and individual rights to privacy. These will be discussed later in the chapter on "Information Technology and Society". Others are service oriented, an example of which is found in the plans for extended networks.

In the United States it is likely that in the early 1980s one or all of IBM's, Comsat's and Aetna Life and Casualty Co.'s Satellite Business Systems (SBS), Xerox Telecommunications Network (XTEN) and Bell's Advanced Communications Service (ACS) will use satellites to transmit data, facsimile and other forms of electronic mail, audit and accounting services together with audio and video conferencing facilities. Transmission speeds are expected to be 50 to 100 pages of material per minute and hence promise to be particularly attractive to business and industry.

These extended networks plan to interface with the various communications modes and employ satellites, fibre optics and other ex-

---

[6] The number of "parking places" for geostationary satellites is limited and could be filled as early as 1983.

isting and emerging technologies. The user will employ the same terminal to send and receive information. Terminals will be connected to rooftop antennas which, in turn, will transmit and receive the data to and from a processing centre with up and down links to a satellite similarly connected to other processing centres and terminals. In this way subscribers will be as effectively interconnected as they are presently by telephone but with all the advantages that sharing a common space can bring. Reports could be edited in real time, blueprints altered and documents exchanged and a percentage of business travel eliminated.

Optical fibres

Microelectronics and the digital system it employs were crucial to the development of a powerful communications carrier. The use of optical fibres to transmit information has assured communications experts that they have at their hands a device that offers virtually unlimited bandwidth at almost zero incremental cost. The fibres are glass threads with a 50-micron diameter core, no thicker than a human hair, that are able to carry concentrated beams of light provided by light emitting diodes or semiconductor lasers.[7] In addition to their bandwidth capacity, because of electrical isolation, they are almost immune to noise or interference.

The bandwidth numbers are impressive. Ten thousand times more information can be carried on a glass fibre than on regular telephone copper pair. A cable containing twelve such fibres, properly protected and reinforced, is smaller in circumference than an average finger. Yet such a cable can provide more than 200 television channels as well as the capacity for two-way communications (see Plate 4 which shows a needle and optical fibres).

For a number of years the bandwidth advantages of optical fibres were widely recognised, but difficulties persisted. These included con-

---

[7] Laser (light amplification by stimulated emission of radiation) is a source of light in which the atoms radiate in step with one another and in the same direction producing "coherent light" and hence are capable of transmitting information at extremely high speeds.

cerns regarding the technology's ability to maintain a signal without significant losses, the questionable lifetime of lasers as a light source and cost effectiveness, particularly as compared with the traditional paired copper wires. As long as messages were transmitted in the analogue mode, glass fibres were effectively hobbled. With the move to digital, however, the advantages began to appear. Since their energy loss is less, light pulses require fewer repeaters than other electromagnetic signals. Glass fibres are not only more durable than copper and not subject to corrosion, but also given a plastic coating they are as resilient as steel. In addition, they provide more security for the user than copper since, if tapped, they must be cut before being spliced, an action which is readily detectable. Finally, the lifetime of injection lasers has improved a million-fold.

While individual fibres are still more expensive than a copper pair, their carrying capacity is such that given the need for additional capacity the advantage lies with glass. A forecast of price with outside limits for margin of error is given in Fig. 11, and a comparison of total system

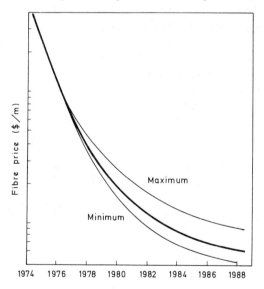

Fig. 11. Cost per metre of optical fibre. Source: Department of Communications, Canada.

costs for fibre optic and paired cable carriers for a given capacity is shown in Fig. 12. The advantage of fibre optics when compared with digital radio is shown in Fig. 13.

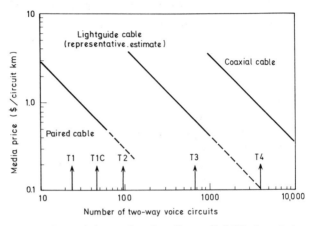

Fig. 12. Transmission media prices. Source: Bell Telephone Labs.

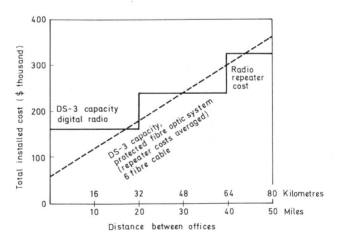

Fig. 13. A cost comparison between digital radio and fibre optics for a given capacity. Source: Murthy, B.N.R., *Gnostics Concepts,* IEEE, 1979.

Reasonably short communication links using this new technology are in operation in various parts of the world, but the decision that convincingly moved fibre optics from an experimental to an accepted mode was made by the Province of Saskatchewan in Canada when orders were placed with Northern Telecom for a 3200-km fibre cable system that would effectively link most of that province's communities. Much longer links, including a transatlantic cable, are now under consideration.

The bandwidth and interactive capacity of fibre optics have spawned two developments which are intriguing in their scope and the promise they hold for future services. In Canada, at Elie, Manitoba, the Manitoba Telephone System, in co-operation with other Canadian carriers and the Federal Government's Department of Communications, is presently installing a system in a predominantly rural area which will provide a variety of services including computerised medical and police signalling, automatic burglar alarm monitoring, remote gas, water and electricity meter reading, energy consumption management, information services, multichannel television and AM/FM monaural and stereo channels. In Japan a versatile experiment is providing 5000 houses in Higashi Ikoma New Town with similar services at an annual cost of 6 billion yen. It is interesting to note that early results indicate that the interactive channels were preferred by four to one over the more traditional fare.

A variety of other applications are being examined by a number of interests including computer, automobile, aircraft, electric utility companies and the military. The small size, light weight, flexibility, strength and reliability of the fibres make them excellent conduits for electronic control purposes. As microprocessors take on more and more functions, the connecting communications links are likely to follow the optical route.

Videotex

Videotex is a new service which uses microelectronic technology to store, encode and decode information aimed at both business and the home. Developed in Great Britain, it soon attracted worldwide attention.

It is possible that videotex could be the most pervasive and socially relevant technology of the 1980s. Essentially it is a source of information stored in a computer's memory which through telecommunications links can be accessed at the command of the user. Its attraction lies in the number of services it is capable of supplying. Virtually all of those listed for the Elie and Higashi trials can be offered with videotex. In addition, a large number of interactive functions can be included such as library services, computer aided learning systems, computer banking, electronic mail, teleshopping, home and factory energy management plus the opportunity to work at remote stations.

Since videotex is a computer based technology, it is critically important that standard methods of describing and distributing the information be adopted or else the social advantages associated with ready and uncomplicated access to information in the public domain will be missed. Standards and protocols must address page formats relating to the preparation of pages for data bases and their transmission to user terminals.

Since these terminals consist of microelectronic components, the degree of sophistication varies according to needs.[8] Nevertheless, it is important to ensure that access procedures be the same for all data bases. It is also important that the terminals be designed in such a way that with some additional equipment they can be used to input information as well as interact with large established data bases. One of the interesting features of the technology lies in its potential to enable the individual of modest means to find a way of distributing material of his or her own without facing the enormous publishing costs and editorial restrictions which characterise existing societies.

The system is designed to transmit information by any or all of broadcast, cable and telephone. Using the vertical interval, that is the unused lines of the television picture, hundreds of pages can be broad-

---

[8] A basic level terminal consists of a decoder, an input circuit, a microprocessor, a 1-K byte random access memory (RAM) input buffer and program variables, a 6-K byte program read only memory (ROM), a page RAM, a character cell driver circuit, a 2-K byte ROM block graphic and alphanumeric shape, and an RF (radio frequency) or RGB (red green blue) driver. Costs for common elements in 1980 are $800; by 1983 they should fall to less than $60.

cast with the aid of an encoder at the transmission end and a decoder at the receiver. Although more costly than broadcast, the additional bandwidth available with multichannel coaxial cable enables a much larger quantity of material to be carried (between 5000 to 10,000 pages per channel). The present switched telephone system allows the user, in addition, to interact with the material. Since the material is stored in computer memory banks, the capacity is virtually unlimited, even if the number of telephone lines is not.

The International Consultative Committee of the International Telecommunications Union have recommended three systems as world standards. They are the British "Prestel", the French "Antiope" (or "Teletel") and the Canadian "Telidon". Prestel, after six months of operation, had 1600 subscribers and optimistically forecasts 100,000 by 1981. Antiope have or are in the process of contracting for 250,000 sets to be provided at little or no cost to replace the conventional telephone directory and "yellow pages". They predict that if the initial experience is encouraging, they will have 30,000,000 sets installed by 1992. One broadcasting network in the United States presently has an application before the regulatory commission for a licence to experiment using Antiope. Telidon is an alpha-geometric[9] rather than an alpha-mosaic system, and while initially marginally more expensive, its superior graphic capability is such that the Canadian Government is promoting it as a second generation system available today. It began on-air transmission in Ontario in January 1980, while cable and telephone systems should be in operation by 1981.

The interest shown in such systems by a variety of countries is great. The Federal Republic of Germany, Holland, Sweden, Switzerland, Japan, the United States and Venezuela are among the nations presently committed to videotex tests based on one or more of the three systems described above.

---

[9] An alpha-geometric coding system permits graphic information to be stored as vectors (lines and arcs) on a conceptual grid of, say, 4096 lines by 4096 columns.

Microelectronics and Information

The information industry

It is not only transmission systems that are important. Information has become a major industry in its own right. Many millions of books, documents and pamphlets are kept in the predominantly print libraries of the world. This represents information that has been catalogued and stored and is only a small fraction of that processed every day.

It would be possible, for example, for the contents of the U.S. Library of Congress to be stored in fifteen to twenty high capacity computers. Already a decision has been made to replace the card catalogue with an "on-line" access system. There is a growing amount of material retained in computerised data banks. By the beginning of 1980 it was estimated that there were 525 specialised electronic libraries containing about 70 million items. The fields covered include education, medicine, law and environment. The costs of using them range from $8 to $125 an hour. The charges depend largely on the complexity of the material that is accessed. Medical information is costly to assemble and requires the time of highly qualified individuals. It is therefore more expensive to use than, for example, biographical data. The prices quoted reflect the philosophy associated with "market economies". Other systems might decide to distribute all or some of the information as a social good with costs charged to the State.

Two of the problems faced by such systems are the difficulty of accessing them and the complexity of many of the indexing systems. Standards and protocols are as badly needed here as in other areas. However, the widespread use of videotex terminals in the home ought to lead to greater efforts to ensure a mass market for material, and this implies ready access. If common videotex standards and protocols are adopted, then information providers will be encouraged to provide their material in an appropriate format.

A difficulty that is not so easily resolved is that of selection. It is expensive to browse through data bases, and the traditional decision tree form of interaction does little to aid the information-seeker in judging the quality of the material available. While some form of rating system would be helpful, the question of who should be responsible for the process is difficult and raises a host of concerns related to ideology, scholarship, relevance, readability and the bias of the assessor.

It would be incorrect to leave the impression that the information industry consists only of bureaucracies and data banks. Any description which omits publishing, broadcasting, education and advertising would be misleading. These will be covered elsewhere, but the technology of microelectronics cannot be divorced from the diversity of functions associated with electronic mail, electronic funds transfer systems and versions of the electronic newspaper. These are sometimes referred to as multiple technologies, since they complement each other, and an investment in one leads to additional commitments to the others.

Word processing and related technologies

Siemens in the Federal Republic of Germany have estimated that 40 per cent of all office work can be automated. Other studies confirm these findings. At a seminar convened by the Science Council of Canada, representatives of government, management and labour agreed that at least a degree of temporary unemployment in the service sector was inevitable.

Word processors are the technical devices primarily responsible. The market for them in the United States alone in 1980 should exceed $1 billion. While existing machines print at speeds ranging from 5 to 100 lines per minute, a new breed of printer can turn out copy at *18,000* lines per minute. The implications of word processing go beyond the conventional typing and dictation functions. Ragan Information Systems claim they can replace 250 filing cabinets with a singular modular terminal for the same cost as the cabinets.

The relationship between interconnected word processor and filing capability underlines the increasing impact computer technology is having on the predicted automated office of the future. Depending on the nature of the software, computers can provide for dictation, copying, duplication, data processing, electronic mail, facsimile, data communication, micrographics and data base management.

Word processors are built using microcomputers with some design elements in the hardware system to make them more user compatible. The American Standard Code for Information Interchange (ASCII) keyboard has been modified to make it almost identical to that of the standard typewriter, and special high level languages have been devised to facilitate an easy transition from conventional practices. The word processor filing system also conforms to the traditional methods of indexing materials. Most companies today use systems built around microprocessors such as the Intel 8080 which possess a built-in random access memory (RAM) to provide the intelligence to make them user specific with external storage accommodated on disks or drums.

Compatibility among various models is a major problem here as in other areas. At the present time some can communicate at the first level, that is they can interact using the teletype (telex) terminal format. At the second level, where different data formats are desired, special software interfaces can be written, but even these are not able to ensure that some of the transmitted material is not lost. The desire on the part of manufacturers for a unique system which could possibly give them a major market share is a principal reason why designers are inhibited from agreeing on overall standards and protocols.

There is a growing demand for integrated systems, but again lack of compatibility has been the most serious stumbling block. Three or four companies in Europe and North America have attempted to develop interfaces with data processing, filing, micrographics, laser printing and photocomposition, but no one has yet emerged as the acknowledged leader. The future will depend upon not only appropriate hardware but also on software to enable the average worker to access, create and manipulate files as well as participate in a growing number of shared systems.

Electronic mail is a function attractive to industry and individuals alike. Reference was made earlier in the section on satellites to "extended networks". The terminals proposed for such networks ought to be compatible with word and data processing equipment to avoid redundancy. Videotex terminals, while initially designed for the home market, have many business and industry applications including electronic mail. Private networks are also being established. The I.P. Sharp "Mailbox" began as a method of sending internal correspondence between offices in Europe, North America and Japan. Other similar networks, such as Comet, Datapac, Tymnet and Telenet, have been developed, and each has made arrangements for subscribers to access the other systems.

Facsimile transmission, relying on telephone lines connected to photocopiers by acoustic couplers, has been available for a number of years, but development was hampered by the relatively slow transmission speeds. While rapid transmission is now possible, it is likely that the service will become but one element within the more sophisticated systems envisaged by the extended networks. The need for rationalisation among the various approaches to the electronic main goal is great if advantages associated with the new technologies are to be realised.

Electronic funds transfer systems

Increasingly banks and other financial institutions with large cash flows have been making use of computer and communications technologies. It is easy to see why. As Martin in his book *The Wired City* pointed out in 1978, the cost of using a credit card is close to 50 cents a transaction. An equivalent electronic funds transfer system (EFTS) is estimated at 7 cents. In the United States about 84 million cheques are written each day totalling $548 billion. If EFTS were substituted, the transaction time would be virtually eliminated and the "float" on accounts receivable correspondingly reduced.

A system consists of access devices such as cards and identity codes, electronic terminals, including cash dispensing devices, and automated clearing houses. Credit cards have been a feature of industrialised societies for the past fifteen years. What is new is their growing use in more sophisticated money management systems.

Electronic terminals, based on advances in microelectronics, take a number of forms. The first were introduced in the late 1960s and were primarily used in the banks as data processing devices. More recently consumer-bank communications devices were introduced and found their way to the exterior walls of financial institutions for use during non-banking hours. Now commonly referred to as automated teller machines (ATM), they dispense cash, accept deposits, transfer funds among related accounts, provide credit card advances and comply with bill payment instructions. They are no longer only on the outside of banks but are finding their way into the offices of large business institutions.

Point of sale (POS) terminals are also on the increase. These are located in retail outlets; they are "on line", that is connected directly with the customer's bank account, and enable the customer to transfer funds directly to the merchant.

Transfer systems require connections not only between the user and his or her bank account, but also need to ensure that similar communication links exist among the various banks. These are known as automated clearing houses (ACH) and are the electronic age counterpart of the cheque clearing function.

With the hardware and the software available from such major multinationals as IBM, Olivetti and Philips, a number of countries have made commitments to the system. For example, France, Belgium, the United States and Canada have already taken action to install POS terminals as well as inter-branch banking. The Federal Republic of Germany's Central Credit Committee has also established a working group to draw up specifications and guidelines for a standardised bank card system.

Continued growth is likely. The United States Congress indicated that there were significant consumer as well as institutional benefits associated with electronic funds transfer systems (EFTS), although they expressed concern that "... the unique characteristics of these systems make the application of existing consumer protection laws unclear leaving the rights and liabilities of users ... undefined." Regulation E, Electronic Funds Transfer 1980, of the Board of Governors of the Federal Reserve System, from which the above was quoted, was designed to react to that concern.

With the advent of videotex terminals in the home and the generalisation of such systems, the terms relating to transactions need to be explicit. The move to the cashless society seems inevitable given the technological push provided by microelectronics and significant cost advantages associated with the transfer of funds electronically.

Personal computers

Personal computers are general-purpose microprocessor based systems. Some models can be plugged in and operated without the need for sophisticated programming. They have enjoyed an enormous growth rate, particularly in the United States. Originally introduced to sell to the hobbyist, their applications have now been such that interest has extended far beyond the relatively limited market originally envisaged for them. As of 1979, approximately 250,000 had been sold and an equivalent number were expected to be marketed successfully in 1980.

The first home computer, the ALTAIR 8800, was featured in a cover story of the January 1975 issues of the *Popular Mechanics* magazine. It was made to sell in kit form for under $400 and boasted that it could out-perform ENIAC. It probably could, for it had a larger memory and was twenty times as fast.

From 1975 to the present the diffusion rate and the increase in sophistication have been phenomenal. Magazines devoted to small systems have proliferated. The largest and most successful so far, *BYTE,* is published monthly and contains between 300 and 400 pages crammed with advertisements for personal computers and almost every conceivable product associated with them. It is available on microfilm both in North America and Europe.

Costs vary with the sophistication of the system. In 1980 a beginner could acquire one in the $500 range to include a 4 K random access memory (RAM), a keyboard, video display and a tape recorder for external programme storage. On the other hand, for prices from ten to twenty times that amount the buyer could get a premium machine with a wide variety of applications. One model, which sold for $4500, advertised a computer able to address up to 128 K bytes of memory with a supplementary built-in 5-inch floppy disk drive, an electric typewriter style keyboard, a silent 80-column dot matrix printer,[10] a black and white monitor and software programmes for handling finan-

---

[10] A matrix printer is a printer in which each character takes the form of a pattern of dots produced by a stylus or number of styli moving over the surface of the paper. Other printers, such as the "Daisy Wheel", use interchangeable character wheels and carbon multi-strike ribbon to provide electric typewriter quality print.

cial and general data. It was also designed to handle multiple languages with PASCAL and BASIC as built-in features. For an additional $3500 the customer could acquire a word processing package which would feature appropriate software, a letter quality printer, an extra disk drive and a high quality monitor. A comparison of fifteen popular models is shown in the following table.

While initial attention may have been attracted by computer games and simple data processing applications, current usage has moved far beyond this level. Many owners are now "on-line" with one or more of the networks described earlier, and hence have access to electronic message delivery systems. The range of interests is reflected in the table of contents of *BYTE*. Examples include: "A first look at graph theory applications", "A computer controlled wood stove", "A financial analysis program" and "Solving problems using variable terrain". Personal computers will add their contribution to the multiplying effect resulting from the merging technologies of word and data processors and videotex.

Microelectronics and Automation

Computer assisted design and manufacturing (CAD/CAM)

The combination of developments in communications and microelectronic technologies has strongly influenced the growth rate of automated manufacturing techniques. Computer assisted design (CAD) and computer assisted manufacturing (CAM) have had a major influence in accelerating this development.

Trial and error have always been fundamental to innovation. Using the computer and an associated graphic display, the time taken to conduct a trial, carry out tests and make corrections where necessary have been dramatically reduced.

The principles involved are relatively simple. A set of picture description instructions (PDI) are fed into the computer and the results displayed on a high resolution monitor. The design is easy to change or modify and, once established, a variety of elements within the design can be tested electronically. As soon as the design is

Personal Computer Reports

Type	Memory	Base list price (US$) 1980 (processor-memory)	Principal programme language	Primary uses	Number installed
Apple II	16KB-48KB RAM; 116KB-16MB d	1195 — 1495	BASIC, PASCAL	C/G, P/O, S/B, E, H/H	55,000
Atari 400 & 800	8KB-16KB (48KB) RAM; 92KB-368KB d	999.95 — 1624.96	BASIC, assembler	C/G, P/O, E, H/H, S/B(800)	N/A
Commodore PET	8KB-32KB RAM; 360KB-1.1MB d	795 — 1295	BASIC	H/H, WordPro, E, P	107,000
Compucolor II	8KB-32KB RAM; 51KB-204KB d	1895 — 2495	BASIC, FORTRAN	C/G, H, P, S/B, E	4000
Cromemco Z-2	4KB-64KB RAM; 92KB-1MB d	1290 — 2780	BASIC, FORTRAN IV, COBOL, Macro-Assembler	H/M, P/O, S/B, C/A	N/A
Exidy Sorcerer	16KB-48KB RAM; 616Kb d	1295 — 1495	BASIC, FORTRAN-80, COBOL-80	E, S/B, H/H	10,000

Personal Computer Reports (contd.)

Type	Memory	Base list price (US$) 1980 (processor-memory)	Principlal programme language	Primary uses	Number installed
Heath H8 & WH8	8KB-56KB RAM; 100KB-300KB d	474 — 1529	BASIC	H/H, E, S/B	12,000
Heath H88/H89	16KB-48KB RAM; 100KB-d	1295 — 2895	BASIC	H/H, E, S/B	3000
Hewlett-Packard 85	16KB-32KB RAM; 195KB-210 tape	3250 (complete system)	BASIC	S/E, P, S/B	N/A
Mattel Intellivision	18KB RAM	300 — 800	NONE	H, E	2000
North Star Horizon	32Kb-64KB RAM; 180Kb-144MB floppy d	2695 — 4330	BASIC, PASCAL	E, H/H, S/B	112,000
Ohio Scientific Challenger I Series	4KB-32KB RAM; 80KB-160KB d	279 — 995	BASIC	E, H/H	8000

Ohio Scientific Challenger II Series	8KB-48KB RAM; 80KB-74MB d/disk	698 — 1695	BASIC	E, H/H, S, S/B	4000
Radio Shack TRS-80 Model 1	4KB-48KB RAM; 55Kb-310KB d	499 — 1446	BASIC, FORTRAN	H/H, E, P,	150,000
Texas Instruments 99/4 (Home Computer)	16KB RAM; 90KB-270KB d	1150	BASIC	C/G, P/O, E, H, S/B	N/A

Abbreviations: KB, kilobytes; MB, megabytes; RAM, random access memory; d, diskette; C/G, colour graphics applications; P/O, professional offices; P, professional; E, education; H/H, home/hobby; H/M, household management; H, home; C/A, control applications; S, scientific; S/E, scientific engineering; N/A, not available.

Source: Datapro Research Corporation, May 1980.

deemed to be acceptable, the computer relates this to the appropriate specifications for the equipment required and retains a control function over the ensuing process.

In February 1980 the *New Scientist* reported on the automated system at the Baker Perkins' factory in Britain. Engineers use information from technical files to create the desired design. The machine shop manager indicated that about half the 10,000 drawings the factory produces annually are made with the aid of CAD. Furthermore, the time taken to plan a new process had been reduced from three weeks to three days.

Automated control of the manufacturing process is not new. Numerically coded paper tapes have been used for years to provide instructions to machine tools for repetitive functions. With the development of CAD, however, it has become possible for these tapes to be generated by the computer as part of the design process. More and more machines are operated under directions from the computer. Microcomputers allow for the testing function to be directly associated with the assembly tools themselves. Usually the control computer is connected to a small television camera which examines the part in question and in turn relates the required direction to the machine tool itself, which contains a microprocessor, and modifies its performance accordingly. Thus faults can be found before they become part of the end product. The computer is also able to manage inventory, shop scheduling capacity and material requirements, purchasing, receiving and product documentation.

The adoption of the techniques described above is accelerating and is becoming an important element in the automation of factories through the use of robots.

Robots

Robotics, a term derived from the Czech word *robota* meaning work, has come into common usage to describe the science concerned with programmable machines designed to carry out a variety of tasks. Early developmental work was carried out in Norway. However, it was the U.S. firm Unimation which first introduced robots in the early 1960s. It may still be the largest manufacturer, although it is facing increasing competition from companies in Sweden, Japan, the Soviet Union, Italy and the Federal Republic of Germany.

Most robots consist of mechanical arms and hands which are able to move in response to instructions provided by microprocessors or computers. Recent developments have led to second generation robots which are equipped with sensing devices to permit a degree of intelligence and through the necessary information feedback to the control mechanism enabled to modify original instructions or indeed create new additional ones.

Small, solid state television cameras supply a visual capability. Typical are those found in the Automated Body Institute of the Ford Motor Company and others being developed by the Swedish firm ASEA. Tactile sensors have also been created. ASEA is also working on units to aid robotic controlled grinding and polishing operations. Although not yet in operation, a prosthetic hand was designed by the Yugoslav designer Tomovic which incorporated sensors in its fingertips enabling it to feel objects. CANSCAN was designed in Canada to probe the miles of tubing in a nuclear powered reactor. A pneumatically operated "tube walker" probes the areas to detect defects, their nature and location.

The number of robots in use is not known to any degree of accuracy. While over 20,000 programmable[11] machines are in operation, with at least 14,000 in Japan, only a limited number are as yet equipped with a degree of machine intelligence. Robotics is an expanding industry, and the major manufacturers are having difficulty in responding to the increasing demand.

Robots are believed to be cost effective and their record of reliability is high. One report showed "up time" for robots to be approximately 98 per cent compared to 90 per cent for machine tools. In the automobile industry, which has become a major user, productivity has increased several times with their general introduction. This is evident in the experience of Fiat, Volvo, Volkswagen, Mercedes Benz,

---

[11] Numbers reported as of May 1980 by the Robot Institute of America include: Japan 14,000; United States 3255; Federal Republic of Germany 850; Sweden 570; Poland 360; United Kingdom 185; Norway 170, Finland 110; Belgium 13. It should be noted that additional robots exist which are not included in this list.

Renault, British Leyland, Datsun and Toyota. They are also safer, less subject to accident and are able to perform tasks too dangerous for human beings. Applications include arc and spot welding, spray painting of both large and small objects and press work for bending metals and pipes. In addition, they are playing an increasingly important role in the space industry. For example, a remotely controlled manipulating arm manufactured by Spar Aerospace is designed to launch satellites into orbit from NASA's space shuttle.

Research and development activity is high. The Swedish Centre for Working Life has constructed a model of a fully automated steel plant. Japan, the acknowledged leader in the use of robots, has established a special Methodology for Unmanned Manufacturing (MUM) project. The Institute for Production Engineering and Automation in Stuttgart is also active in the field. In the United States one of the leaders is the Institute for Industrial Automation at Stanford University. The Massachusetts Institute of Technology (MIT) is also active in the field of pure research in this area.

Some of the constraints presently faced in the use of existing industrial robots include their ability to cost effectively handle batch sizes of less than 1000 units. Numerically controlled machines still enjoy a considerable advantage in this area, since they are suitable for work on a limited number of pieces. The rapid growth in the sophistication of microprocessors and the ability of the newer bubble memories to operate in an atmosphere which is far from clean is increasing the flexibility of the robot and will undoubtedly soon enable it to economically deal with smaller and smaller batches.

Another limiting factor is the conservatism of many industrial managers and their tendency to try to match the use of the robot to the traditional manufacturing process rather than to re-examine the process to see how it might be modified to take advantage of the capability of the robot to perform tasks which require precision, patience, continuity and reliability. The lack of standards is also a problem and results in unnecessarily high costs. Standard and modular hardware accompanied by appropriate software packages would not only lower prices for the machines but also increase significantly their applications.

New and interesting developments are taking place. In the Soviet Union, using three-dimensional models of the environment, sensors and feedback techniques, mobile robots have been created which are able to pick their way through typically cluttered environments. The Japanese have been extending their work in visual sensors and have now developed one which is able to distinguish up to ten colours. In the United States at one company, significant work is being done on the development of flexible manufacturing systems, where the process is architected to provide for increased individuality and at the same time more complementarity. In Italy the Olivetti Sigma assembly robots have solved assembly problems where rigid automation was inapplicable.

Microelectronics as a substitute for mechanical and electromechanical devices

Since integrated circuits require only miniscule power units, they are challenging many conventional mechanical and electromechanical devices. The impact of microelectronics on the watch industry is a classic example of the dramatic changes effected by this relatively new technolgoy. Before the Second World War 90 per cent of all watches were made in Switzerland. By 1979 this percentage had been cut in three and employment in two. This occurred despite the fact that the Swiss themselves had developed one of the first prototypes of the electronic quartz watch as early as 1968. Perhaps because of the success they have achieved using traditional methods, the industry failed to adjust to the new technology. Not so in Japan, where the watch industry rapidly converted to the electronic mode and as a result gained a large share of the market. Belatedly the Swiss recognised the need for a radical change, and a major manufacturer (Omega, Tissot and Hamilton) made the necessary adjustments. It acquired integrated circuits from Philips and developed and manufactured quartz crystals within its own factories. Now over 50 per cent of its output consists of electronic watches.

Electronic games and pocket calculators are commonplace today, but perhaps the largest user of microprocessors and microcomputers is the world's automobile industry. Experiments related to emission control and fuel economy go back as far as 1969, but the efforts were fragmented and minor in their impact. Today these electronic devices are used to measure temperature, engine speed, pressure and oxygen content of exhaust gases. This information is stored and processed to adjust spark timing and choke control. Volkswagen has installed a completely electronic ignition system. Microelectronic chips have also been used in anti-skid braking systems and digital dashboards which feature trip computers measuring fuel consumption, elapsed time, estimated time of arrival and average speed achieved. In addition, some models contain electronically controlled suspension systems, door locks, radio antennas and power seats. The value of semi-conductor components sold for cars and their major applications are shown in Figs. 14a and b.

Other consumer goods which contain integrated circuits and which are, or shortly will be, available include microprocessors on a card and "smart" phones. The first are similar to credit cards except that they contain an erasable memory which indicates an amount of credit prepaid by the user. In its application it can be compared to a certified cheque, the face value of which is automatically diminished as goods

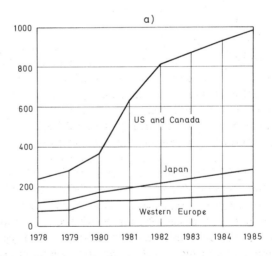

Fig. 14a. Millions of dollars worth of semiconductor components sold for cars. Source: Dataquest Inc., B.W. est, October 1979.

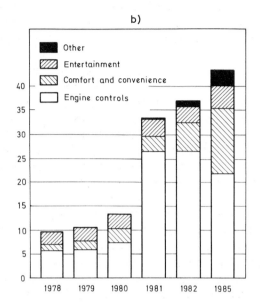

Fig. 14b. Dollar value of semiconductors per car. Source: Dataquest Inc., B.W. est, October 1979.

or services are purchased. One, an Italian device, called the x-card contains a 136-bit memory with added security and control functions. Another, the CII developed by Honeywell-Bull and licensed by NCR, Siemens and Schlumberger, has a larger capacity than the x-card and contains both a memory and microprocessor. With the cost of misuse of conventional credit cards increasing, the potential of the new "pay in advance" approach is becoming increasingly attractive.

"Smart" phones are connected to intelligent switching systems. For example, the digital crosspoint (DX) chip is a basic component in Mitel's SX-2000 system. The DX provides digital voice or switching data on 256 channels and represents that company's first very-large-scale integration (VLSI) design. A square with about ¼ of an inch per side, it has over 36,000 transistors. Random logic, shift registers and static random access memory (RAM) circuitry are all utilised on the chip. Smaller systems are also available for offices or homes. The

SX-10 from Mitel is the world's first talking switching system and can handle up to sixteen internal phones and eight external lines. The 25 × 13 inch printed circuit uses electronically digitalised words assembled in short phrases under the control of the system's microprocessor. Any language, or dialects of a language, can be used with a change to any other accomplished by the replacement of a memory chip.

These phones can be programmed to act as alarm clocks, regulate temperature, provide a different ring to distinguish internal from external calls, remember numbers which, when dialled, encounter a busy signal and then recall when the desired line is open and remember precoded numbers and dial them upon request given a one or two-digit signal. They are capable of determining the least cost routing where options are available to the user, of eliminating the need for an operator and of acting as an answering device when required.

Microelectronics have been substituted for a number of mechanical and electrical devices. Some examples have already been mentioned. These included word processors, videotex, electronic funds transfer systems, watches, telephone exchanges, telephones and a number of control devices which improve automobile safety and performance. However, there are many other instances. For example, Telex systems have largely replaced the telegraph and now permit the user to enter messages into a central computer which either forwards them directly or, if the receiving machine is busy, will keep calling until completion. Electronic typewriters were the forerunners of word processors and replaced many electric typewriters as a result of their ability to store standard paragraphs and reproduce standard letters. Electronic taxi meters are replacing mechanical meters and at the same time provide computer controlled communications with a central dispatcher. Microcomputers are also being sold as energy savers. With thermostats they monitor temperature and humidity and automatically adjust furnaces, air conditioners and humidifiers according to a predetermined programme. Microelectronic devices now control the aperture, focus and shutter speeds of cameras. These are only a few of the applications which might have been described. The use of microcomputers and microprocessors in mass transit systems and in medicine is covered in the following sections.

Tele-information for transit systems

A computer has effectively demonstrated its ability to increase both traffic volume and cost effectiveness of public transit systems during a two-year trial in suburban Toronto. Buses are tracked electronically by a central computer linked to a voice synthesiser and interfaced with the telephone system. Any caller wishing information dials a given three-digit number plus a two digit number representing the bus route and a further two-digit number representing the desired bus stop. The response is immediate and arrival times are given for the next two buses. After the initial year an increase of 23 per cent in the number of passengers for the given routes was recorded of which a minimum of 12 per cent was attributed to the "on-line" computerised system. Not only is the system cost effective, it also has significant implications for energy conservation. The success of the experiment has led other centres to install similar systems. In the future videotex could eliminate the need for even the telephone interface.

Microelectronics and medicine

Computers and associated communications media have been used as medical aids for over a decade. Remote diagnosis and treatment were an early feature of communications satellite experiments, and in most cases justified the expectations held for them. In all likelihood this application will expand as direct broadcast satellites increase in number and ground receiving stations decrease in price.

Interactive data bases have also provided the profession with easy access to hundreds of thousands of bibliographic citations and an increasing number of detailed descriptive outlines. More recently, electronic aids to diagnosis have formed integral parts of specially designed software programmes. In one instance more than 3000 symptoms have been described and related to some 500 diseases. In another, the linguistic behaviour of paranoids have been simulated as an aid to treatment.

Belzer, in an article in the January 1980 edition of *Information World,* described an experiment which generated widespread interest. It involved the use of electronic stimulation to initiate the regeneration of a rat's amputated limb. Human broken bones which had hitherto refused to mend began to do so as the result of a similar process.

Future Developments

Supercomputers

It has been predicted that the large computers of today, which include hundreds of thousands of logic gates and 32 million bit random access memory (RAM), will by the end of the century be able to be contained in a shoe box and sell for $1000. If this is to happen, then certain existing limitations will have to be overcome. These include the speed at which the switches operate and the size of the individual components. Progress is being made in both areas as well as a new appreciation of the application of infrared light as a conductor of information in digital form.

Efforts include the synthetic metals (electrically conductive organic solids) and gallium arsenide. Silicon on sapphire (SOS) is an attempt to minimise power dissipation and hence reduce the size of the transistors. The most widely known experiments, however, concern what has come to be known as the Josephson junction.

Josephson junctions were named after a Cambridge University graduate student who discovered that a current would flow through a junction between two superconductors.[12] The time taken for the gate to operate is extremely short, 1/7 trillionth of a second, and the superconductors can be packed very closely together. The disadvantage lies in the need to achieve a temperature of close to absolute zero in order to provide for the high degree of superconductivity. This implies immersing the units in liquid helium, not an inexpensive process. Nevertheless, should it result in a supercomputer which can carry out almost 100 million instructions per second, the costs could be justified.

Competing with the Josephson junction superconductor principle is the "optical computer". For the switching function two stable states of transmission are required. This bistability is achieved by an optical resonator known as a Fabry-Perot interferometer, which reflects light between two parallel plates to achieve the desired effect. Its pro-

---

[12] A superconductor has virtually zero resistance to the flow of an electric current and is achieved by cooling a metal close to absolute zero.

ponents claim that it is as fast as Josephson junctions with the added advantage of compatibility with fibre optics.

Laser switches also promise to improve switching rates. Laser means are used in conjunction with antimony which jumps the power of the beam and creates an incredibly fast switch.

Infrared light was found to be an effective conductor of digitalised data. Since it behaves like ordinary light it can be either aimed or diffused, but will not penetrate an opaque material. Given these characteristics, it has been very effective at short range. The signal is created by a light-emitting diode and received by a photodiode. It was used initially in Europe for remotely controlling radio or television sets. Today it is finding increasingly application in toys and video games, cordless headsets and cordless telephones. IBM is experimenting in its Zurich Laboratories with a room bathed in infrared which, through judicious use of frequencies, is capable of carrying messages back and forth among terminals without the danger of interference. The cordless society may be on the point of arriving.

Other developments

Other developments are almost too numerous to mention. Some of the more important include the use of very small lasers to print or store computer data in a liquid crystal, computer control units and bubble memory controllers.

The first of these, crystallised data, is already being examined as a way to improve computer assisted design techniques and as a device which will not only add to the quality of the image on the picture phone but also enable detailed charts and information tables to be examined without the need for a facsimile transmission.

Magnetic bubble circuits are also available with digital technology. These have enabled a large variety of information to be stored including recorded announcements which require no moving parts.

Holograms and self-programming computers are presently interesting areas for research. The day may not be too far off when video conferencing may be able to present three-dimensional images that will appear to be as real as the original. At the present time, however, the concept is still in the realm of science fiction; the use of holograms as memory storage areas is not, and it promises to be the next major breakthrough in information storage and retrieval technology.

Acknowledgements

The author is grateful for the assistance and advice of scientists and others associated with the Science Council of Canada, Bell-Northern Research, Northern Telecom, Mitel Corporation, Teleride Corporation of Toronto and the GAMMA group of Montreal without whose help much of the data could not have been assembled. In addition, both Bell-Northern Research and Mitel provided some illustrations.

IBM were also helpful in providing information and illustrations of developments related to the art of computation. Much of the early history was garnered from the excellent books edited by B. V. Bowden, *Faster Than Thought,* London, Pitman, 1953, and G. Fleck, *A Computer Perspective,* Cambridge, Harvard U.P., 1973.

Hundreds of articles and publications were consulted. Where direct references were used, the source is indicated in the text. However, various issues of *Scientific American,* particularly those of September 1977 and May 1980, contained information regarding microelectronics, microprocessors and memories, that is fundamental to any work which professes to deal with the field. The *New Scientist* was also a valuable source of information about current developments. The *Electronics Journal* and *Canadian Electronics Journal* were consulted on a regular basis and a number of articles, particularly the April 1979 edition of the former and the March 1980 edition of the latter, contributed useful background material. Other publications which were particularly useful include *The Industrial Robot* of March and June 1980, *Science* of 23 May 1980 and various editions of *IEEE Transactions on Computers* from March to October 1980.

Plate 1. Abacus.   *Reproduced by permission of IBM*

Plate 2. Jacquard's master loom.   *Reproduced by permission of IBM*

Plate 3. ENIAC.

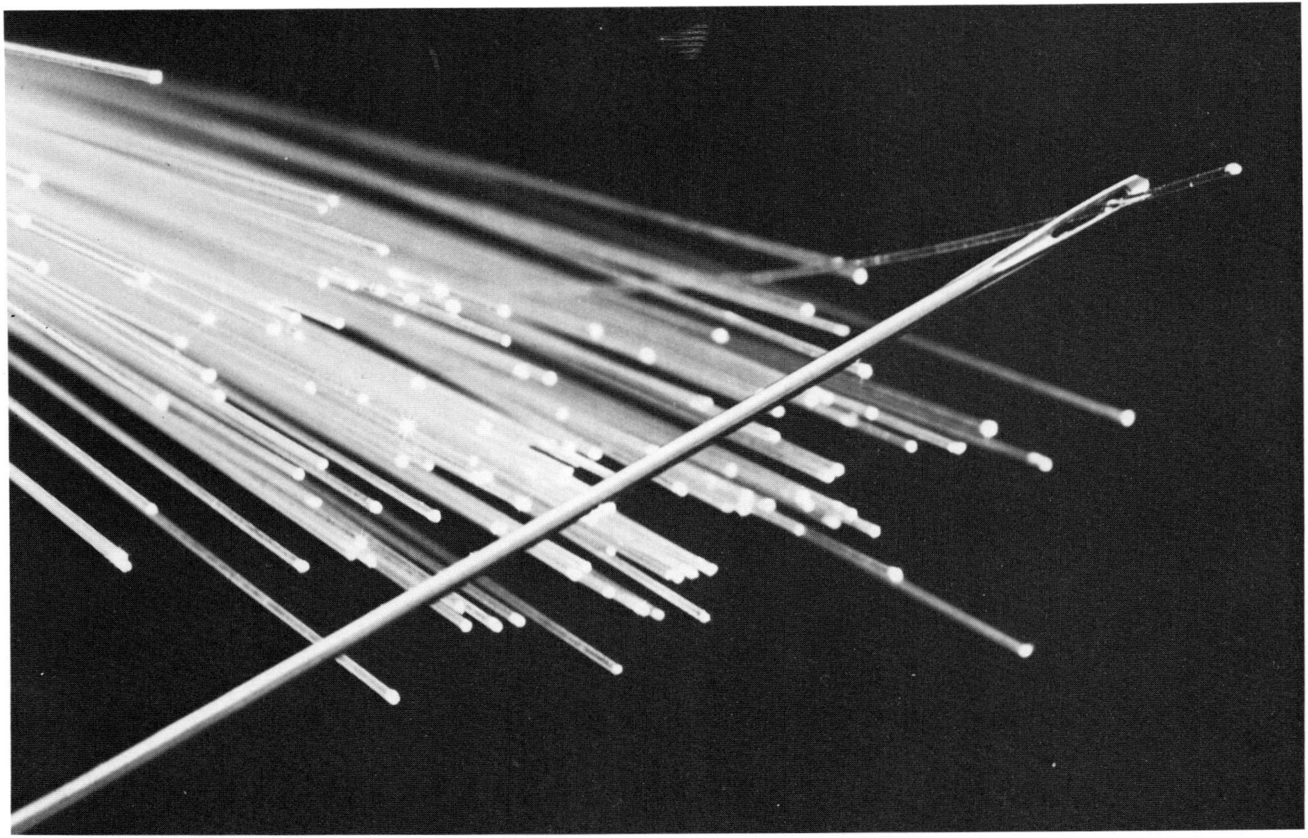

Plate 4. Optical fibres.

Reproduced by permission of Bell Northern Research

3
The Technology Applied

RAY CURNOW AND SUSAN CURRAN

Introduction

There is virtually no field in which microelectronics, and the associated technologies it has made possible, is not being applied today, or will not be applied in the near future. The impact of this new technology is literally spreading everywhere. As we look around the showrooms of new products, and increasingly in the home, the office and the factory, in schools, hospitals and transport systems, the silicon chip is becoming a familiar technological building block. Maybe it is not visible and is often taken for granted in the product leaflet or instructions, but somewhere behind the control knobs or switches there is a shiny sliver of silicon!

The speed at which microelectronics is being applied varies considerably, however, from one functional area to another. Similarly, the availability of new products and processes varies from one country to another. The main factor limiting the pace of application development is not the difficulty of identifying areas in which microelectronics can profitably be applied, but our ability to develop and exploit actual applications.

In addition, the trading policies and economic circumstances of each country will determine when and whether these new technologies and products become readily available.

Development of application areas

At this stage, the prime motive in developing applications is not so much technoeconomic advantage—for microelectronics is proving to have very positive technoeconomic advantages throughout this wide span of potential application areas—but it is based on what is seen as a "needs-imperative". In the military field in particular it is the needs-imperative which provides the force behind the rapid exploitation of this new technology.

An analysis of the pattern of diffusion of earlier, and on the whole much less potent, technologies shows that they are generally exploited first, and to the maximum extent possible, in the military field. The search for increased reliability in electronic systems lay behind the early developments of integrated circuits, and the micro-miniaturisation possible today is but a consequence of that search for reliability. The chapter "Microelectronics in War" explores in detail the applications and use-considerations governing this forced breeding ground for technological development.

One inevitable consequence of this pattern of development is that the most advanced capability in microelectronics and its applications will belong—and indeed, already does belong—to those countries with a high level, both absolute and relative, of military/defensive expenditure, and with a large military-oriented research and development capability. Commercial research and development expenditure also, of course, plays a very significant role, but where the level of research sponsored by the military establishment is high, this will naturally provide a massive multiplier to the results of the civilian research and development programme. Much experimental spadework will have been carried out in this economically protected field, and commercial development can be concentrated in areas where the initial work has indicated the existence and the outline of promising potential applications.

Many microelectronics firms in Silicon Valley in California can be traced back to founders who worked with Fairchild Camera Corporation, a major U.S. defence contractor, and the original concept of the integrated circuit can be traced to the Royal Radar Defence Establishment of the United Kingdom in 1952.

Such positive spin-off areas, which reap the indirect benefits of military research, in the form of commercial application include, for instance, work in aviation medicine and space exploration. This has major consequences in, for example, the recovery of deep-sea divers and in decompression problems associated with tunnelling. Spin-offs can also be identified in the production of materials with completely new and useful properties. Indeed, it is likely that the production of the very specifically "pure" materials required as base materials for future microelectronic development may have to be carried out in sterile, non-magnetic, non-gravitic fields. These conditions are as yet only available in space outside the earth's atmosphere.

We have already, then, identified two factors affecting the spread of microelectronic applications. Firstly, the closeness of the application itself to the military research endeavour, and the extent to which it benefits from positive spin-off; and secondly, the degree of prominence which the individual country maintains overall in the research and development field, both military and commercial.

It is certainly true that the OECD countries, including Japan, appear to be forging ahead much faster in the development and spread of microelectronic applications than is the rest of the world. It may seem paradoxical that Japan, with a defence effort miniscule in relation to that of the United States, should now press hard on that country in microelectronics. This is entirely explicable, though, by the well-known ability of the Japanese to innovate and to exploit modern technology to the full.

Of course, the open scientific publication climate of the United States helps all OECD countries. It is tempting to ascribe the slowness of the CMEA (Council for Mutual Economic Assistance— COMECON) countries in developing commercial applications both to the current rigidity of central planning, and to the apparent communication gap between defence effort and civilian exploitation.

The time scale of development

Any indication of the time scale in which developments are to be anticipated must take into account the individual country under consideration, and its position in this field of technological development. Products and process applications which are already commonplace in some Western countries have yet to appear in many Third World countries, and an indication of what is to be anticipated in the next ten years would be meaningless without consideration of this additional dimension. The economic disparities between countries so evident in the world today are likely to be increased by these developments, forming an additional hindrance to the uniform spread of innovation.

A minority of microelectronic applications, particularly in the computing field, rely heavily on further technological developments in the field of microelectronics—for example, increases in processing speeds, or in the density of integration of components—for their own development. The same is not true, however, of a great many applications in a wide spread of areas. The emphasis here—in the development of microelectronically controlled household goods, education aids and aids for the disabled, for instance—is not on exploiting the results of future research to be anticipated in the next ten years, but on taking fuller advantage of the technology currently available. Development lags behind research, and the gap between the laboratory prototype and the consumer product readily available in the world's department stores and supermarkets is growing steadily wider.

To restate this in slightly different terms, the determining factor in the emerging pattern of microelectronic applications is not what we *can* do, but what we *choose* to do—individually or collectively, but always within the economic limits and mechanisms of our societies. Technically, a vast range of capabilities are already available, and only the minority of technically demanding appllctions even approach, let alone are constrained by, the remaining technical limits.

The real factors affecting the choice of applications for large-scale development are not technical, but economic and social ones. Products and processes which incorporate microelectronics will do so not because of the exciting technical aspects of their design and implementation, but because of the benefits they appear to offer to their consumers, and thus to their producers. The applications which spread fastest are those which prove to be economically attractive.

In this chapter we shall be looking in greater detail at a number of applications of microelectronics. These applications have, to some extent, been chosen to illustrate some of the technical and functional areas in which microelectronics is being put to use. But more importantly, they are illustrations of the ways in which societies with competitive economies are choosing to channel their development effort.

These are products and services for which there is a proven or anticipated demand: and it is that demand, and not the technical originality of their design, which will determine their real success or failure, their widespread adoption or their rapid disappearance. There is no place for a product which is ingeniously designed, which works perfectly, but which nobody wants.

Of course, there is a "producer push" as well as a "market pull". Given the high investment already sunk in research and development, it is inevitable that the semiconductor companies and divisions of equipment companies are working hard to find and develop new markets.

The wider impact

It is not in discrete products alone, however, that microelectronic applications are having an impact. Less tangible effects of the microelectronic revolution can be perceived in the evolving infrastructure, and in patterns of social interaction and lifestyle as a whole. Even an individual product development—the television, for instance—can show widespread "ripple" effects throughout the structure of society. The future with microelectronics may lead to even deeper changes—in where we work, where and how we are educated, how our political systems function, and so on—which all hinge to a greater or lesser extent upon the simple technical developments which underlie the microelectronic industry.

In reviewing some applications of microelectronics and associated technologies in more detail, we shall divide our field into four areas: based, not around technical distinctions, but around the scope of the various impacts which microelectronic applications can be expected to have.

First, we shall look at products, and how microelectronics is being used within them. Secondly, we shall consider microelectronics as it is applied within the production process. Thirdly, we shall review some aspects of the effect upon the economic and social infrastructure, and upon the provision of services of all kinds. And finally, we shall consider how microelectronics is affecting the research and development effort itself. The concluding remarks will attempt to sum up this wide application field, and give a perspective upon it.

Product Applications

In looking at the application of microelectronics in end-products it is important to distinguish between existing products and new or evolving products. In the familiar, already existing product, the use of microelectronics is primarily a technical development: the new microelectronic circuits replace earlier technology and perform essentially the same functions, but more efficiently. In the evolving product, the incorporation of microelectronics into the product leads to a new end-product itself. The product definition evolves, as new possibilities become apparent: and the result is a radically different product, with entirely new capabilities or a totally different mix of capabilities from the previous product generation.

Direct substitution

The direct substitution of microelectronic circuits for older technology—electrical circuits based upon the wiring together of discrete components such as transistors, resistors, capacitors, relays and so on—has, on the whole, preceded the evolution of new products. The existing products have a proven and identified market demand, and little or no consumer resistance need be anticipated to the improved models. Here the powerful technoeconomic benefits of microelectronics can be seen at perhaps their clearest. The microelectronically controlled products are typically more powerful, more reliable, very often smaller and cheaper than their predecessors: and it is little wonder that they are spreading rapidly.

Development has been particularly fast in areas where the scope for improvement in the performance of existing functions was clearly apparent, or where the general evolution of society has placed additional demands upon product performance. A typical example might be the clock and watch industry, where in the space of a few years, the new technology has already taken over half of the worldwide industry. This is a particularly interesting case, since the four successive functional elements of a timepiece: the periodic impulse, the conversion of that periodicity to human time, the display or other human interface (sound, for instance), and the renewal of the periodic impulse through an energy source (for example, a solar cell), have all given way to microelectronic based technology. The transition is 50 per cent complete, and the rate of penetration increasing. Of the 350 million watches to be produced in 1981, 60 per cent will be quartz or oscillator based.

Another example is the automobile industry, where more stringent requirements for the control of pollution from exhaust emissions, and the stress on fuel efficiency created by rocketing oil prices, have forced the pace of development of microelectronic devices to control these functions. There is scope, though less dramatic, for the use of microelectronic controls in washing machines, refrigerators, ovens and other domestic appliances, and many new ranges being brought onto the market in Western industrialised countries boast this technical refinement. And in the office, the photocopier and the vending machine are examples of existing individual products where the level of user satisfaction was relatively low, and where the application of microelectronics to control current functions is leading to improved quality and reliability.

Product evolution

The photocopier also provides a typical example, though, of the "evolving" product. At the lower end of the price range in particular, the application of microelectronics to control the copying functions is leading simply to a better, cheaper photocopier: doing what previous photocopiers did, but more efficiently and effectively. Further up the product range, the development of integrated office systems is leading to the virtual disappearance of the larger, high capacity photocopier as a separate piece of office equipment.

The so-called "intelligent" copier can not only display or tell the user when the supplies of paper or toner are running out, but is self-diagnostic, produces copies where copies are needed, and/or sends electronic copies to other output devices elsewhere. Copiers are taking over decision functions which used to belong to the computer, and the computer and word processor in their turn are taking over the copier's function. The simple act of reproducing an image on a piece of paper has been absorbed into the business of controlling the flow of information within the enterprise. Whether the image is required on paper, or in electronic form; whether a document is already in existence, or needs to be created or altered, are functional questions which the automated office handles in an entirely different way.

A similar process can be seen at work in the development of the pocket calculator industry. The pocket calculator itself is often cited as one of the truly *new* products created through the application of microelectronic technology. Though mechanical, electrical and then electronic calculators existed previously, they were cumbersome and expensive, and so extremely limited in their range of uses. The emergence of the pocket calculator opened up an entirely new market, selling these devices to people who had never previously contemplated buying an electronic calculator.

Today, the pocket calculator has reached a rock-bottom, virtually throwaway level in terms of price. It is a casual buy for many Westerners. And the industry has responded to the challenge to evolve and stay alive, by developing new products which combine the calculator functions with other functions also made possible through microelectronic technology: from the serious (the incorporation of an alarm or stop watch, the expansion of memory capability, the development of the programmable "pocket calculator" which has a full alpha-numeric keyboard, now available at the price of a simple calculator five years ago) to the frivolous (devices for calculating horoscopes, biorhythms, and so on). These are all evolving products: offering new combinations of functions in an attempt to catch the fancy of a fickle, easily saturated market.

The reduction in size made possible by microelectronic circuitry has had much to do with this expansion of capabilities. A new product with four or five distinct, related functions may be even smaller than the single-function electronic product which it is replacing. The small size of microelectronic controls means, too, that they can be used in places where electronic circuitry would formerly have been too cumbersome. Many examples can be found in the paramedical field. Perhaps the most exciting prospect is the development and refinement of devices which transform sensory input—vision or sound—into electronic impulses which can be transmitted to the brain. With the emergence of these microelectronic "eyes" and "ears"—still under development—it will become possible for the blind and deaf to learn to see and understand again.

Many paramedical products now being tested incorporate a degree of "intelligence": and this dimension is one in which an increasing degree of product evolution can be expected. The essence of a computer, as opposed to a calculator, is that it responds to the results of previous operations, without the need for intermediate user input: and the same is coming to be true of many other products using microelectronic intelligence.

In the medical field, these include pacemakers which adapt automatically to changes in the patient's heart condition, and "intelligent" wheelchairs which can be taught to steer themselves. In the home, we are seeing the emergence of environmental control systems, automatically regulating heating and air conditioning in response to feedback on ambient conditions. "Smart" security systems can be taught to respond only to stimuli by authorised users—voice or finger prints, or individual codes, for example. And the home computer is, of course, becoming a familiar product as the price of computing drops to a level where it is well within the reach of many domestic consumers.

The pace of implementation

Just as existing products have, on the whole, converted to microelectronic technology more rapidly than new products have evolved, so the individual, discrete product has proved faster to adapt than the larger integrated system. The telephone system, for instance, represents a massive piece of capital investment, with a relatively long life, and there is no question of switching over an entire national system to microelectronic technology overnight. The adaptation must be gradual, and the newly introduced equipment remain compatible with the old. A number of new types of telephone exchange are being integrated into systems in this way.

Microelectronics-based attachments to, and enhancements of, the telephone system are also proliferating. Among them are answering machines, single button dialling devices, and devices for storing or redirecting incoming calls.

A similar factor also affecting the pace of implementation is the product design and production cycle. For many larger consumer durables—washing machines, televisions, automobiles and so on—the time lag from the planning of a new product to mass production may be considerable. Thus a car has a typical seven-year design phase before manufacture and a washing machine three years.

Microelectronically enhanced versions of these products, microelectronic systems incorporated in these constructs, might have been expected to appear rapidly. In practice, though, the product life-cycle has delayed the impact. Virtually all the designs now on the drawing board, however, envisage using the new technologies.

The ready availability of components also affects the speed of take-up for microelectronic applications. The microprocessor, with its remarkable ability to take on the characteristics of almost any electronic circuit, and to be dedicated, or indeed rededicable, to changing requirements, is likely to become the standard building block of the future for many products. But other critical elemental building blocks such as displays, sensors, actuators, interfaces, keyboards and so on must also be taken into consideration. As the market for such components matures, supply difficulties (a very real problem in recent years) should become a thing of the past, and an increasing degree of standardisation between manufacturers should emerge. The functional requirement in use for many different products is the same, and the standardisation process is already observable in the marketplace today.

Computing power

However wide the variety of products using these components of microelectronic technology, the prime product application area—and the area where the most dramatic advance has occurred in product power, capability and efficiency—must be that of computing products. Not only has the traditional mainframe computer used the extra density of the microchip to the full, but incredible computing power is now available technically from machines costing only a few hundred dollars. Of course the peripherals for such computers are still relatively expensive, since mass storage devices (such as tape or disc) still require moving parts, and so do most printers. Nevertheless, there are already available (but still at a relatively high price, until mass acceptance occurs with the replacement of previous devices) non-impact printers and non-moving mass storage such as bubble memory.

These new types of computing peripherals are based on associated technologies which have only become feasible through the development of microelectronics, and they provide a good illustration of the type of radically new product which microelectronics has brought into being indirectly. Ink-jet printing, where a fixed jet of ink is shaped magnetically to form characters or graphics, owes its existence to the microelectronic controls which are now available, and which make it technically and economically feasible. The laser-xerographic non-impact printing route has also been made feasible in this way. Both technologies offer fount-free, on-line changes of print format, and are infinitely more flexible than traditional printing methods.

Fibre optic transmission, too, is a major new technology made practicable through the use of microelectronic controls. It is already used in situations such as in-house office systems, or of course for substitution as need or traffic demand occurs in telecommunication networks. The use of fibre optics can be expected to extend rapidly from the present experimental developments in automobile wiring harnesses.

Process Applications

Having discussed the types of end-product which are emerging at different rates through the economy—and, of course, at different rates in different societies—it is now important to consider the processes by which these will be manufactured, and the impact of microelectronics on those processes. This, too, will prove to be a major field for the application of microelectronics.

Mechanisation and automation

One aspect of the importance of mass produced goods in general, and mass consumer products in particular, is that these have traditionally been the products of assembly industries. Steady evolution in production methods has led to a fairly high degree of mechanisation, for instance with the development of the conveyor belt. But in the final analysis, the backbone of this sector has been the human assembly process.

The substitution of microelectronic circuits for electronic circuits and other older technologies has brought about a dramatic reduction in the number of sub-assemblies required in the typical assembled product. This in turn opens up the possibility of complete mechanisation, or indeed full automation, of the assembly process. The use of microelectronics also allows for local computing capacity to be available at each stage in the production process, and individually for each part-assembly to be handled. The convergence of these two rapidly increasing capabilities has resulted in full automation spreading throughout such types of engineering manufacture. Already, for example, several car assembly plants are organised with production robots in place, such that different models can be assembled on the same assembly line.

One general point must be made about all these integrated business systems, where production, clerical or warehousing functions are brought under the same, often centralised, hierarchy of control. The major implementation problem is that the present pattern, scale and methods of these operations evolved as a consequence of the previously available technologies, and the act of computerisation may not fit easily into systems designed on a quite different basis. Although fully integrated systems are on the horizon, it is likely that their full and economically optimised implementation will await a major restructuring of business. Thus, it has been argued that common production centres will be shared by a whole sector or branch of the economy, leaving competition between firms to take place in the marketing, support or service areas. Such a scenario may seem far-fetched (though it is common in petrochemicals today), and it is more likely that new entrants, unencumbered by previous plant decisions, will reap the full benefit if they can locate in buoyant economies.

However, the rapid installation of full automation in mass produced products has perhaps tended to overshadow the practical possibility of applying such flexible automation methods to smaller batch processes—processes which, in many cases, have until now been relatively unmechanised. This smaller scale automation has to a large extent been restricted to date to processes with a very high added value, such as computer products or defence products. But as the ability provided by microelectronic technology to reorganise production processes becomes more clearly understood by management, and given the simplification of computer processes which is now well under way, it is likely that the vast bulk of discrete engineering products will be processed in this way, virtually regardless of batch size. There are interesting prospects, too, for the use of microelectronically aided production techniques in quite different industries such as garment and shoe making, where automatic size-changing is revolutionising cutting methods.

Computer aided manufacturing and design

The integration of computer aided manufacturing (CAM) with computer aided design (CAD) is also a growing trend. Here again it has first been introduced in the defence industries, or in the high value-added investment goods, but it may be expected to spread into a wide variety of other areas of manufacture.

Computer aided design is an old term, computer aided manufacturing a more recent one. Clearly CAD-CAM in combination is not only feasible but desirable, the output from the former driving the latter. At present, the information-product based equipment suppliers are swallowing their own medicine, and are amongst the foremost users of CAD-CAM. It is no coincidence that Fujitsu are one of the leading CAD-CAM suppliers with their latest FANUM systems, and the integrated nature of the major Japanese industrial empires lends power to their innovative ability. The application to car assembly has already been noted, but in the field of aerospace, too, a repetitive but high value-added assembly process, CAD-CAM is well established.

The discrete engineering industries, which cover all the mechanical engineering and metal-working industries, are particularly ripe for the full adoption of automation. The effect here will be very visible, as indeed will the opportunities opened up for new products. It is interesting to note that in a much earlier analysis of the impact of automation, that of Einzig in 1957, he drew attention to the disadvantages of heavy capital expenditure, self-feeding obsolescence, reduction in choice and vulnerability to breakdown associated with the automation methods then coming into use. Microelectronics gives that extra flexibility which changes those disadvantages to advantages. Of course, the imagination and reorganisation of the engineering industries to achieve full automation requires much effort, to say nothing of the attendant labour force reorganisation.

Summing up, the factory of the future will be highly automated. Machines will do most of the work and, most important, automation will be flexible. The most advanced manifestation of inflexible automation is the transfer line. A series of machine tools, each controlled by computers, perform machining operations on a part or component as it proceeds along a conveyor belt from one machine to the next. This type of equipment is typically used for mass produced items, and entails heavy investment.

Machining centres are computer numerically controlled machine tools (CNC) that can combine some or all of the four basic types of machine tool—that is, milling, grinding, drilling, and in future turning. They can be equipped with as many as forty different tools. While one tool is working, the following one is automatically set up for the next operation.

Flexible manufacturing systems use machining centres and industrial robots which are linked together in such a way that the parts being worked on can travel from one centre to the other in different sequences under the control of computers. Both the machining centres and the flexible manufacturing systems permit the continuous manufacture of different items in small batches.

Needless to say, important developments are also taking place in the manufacture of mass produced items. The importance of the flexible manufacturing system is that it permits the production of individualised products. The crucial element is the programme that runs the system—in other words, the software. As engineering problems are solved and the economy of computer aided design and draughting increases, a fully integrated business system will be available under which machining operations, stock control, management information and clerical activities are done with computerised equipment.

Continuous process operations

In the continuous process industries—for example, the oil, petrochemical, chemical, plastic and gas sectors—the impact of microelectronics is already being felt through the technical revolution in instrumentation and control. Many new sensors now available can provide a direct electronic reaction to stimuli from changes in the production process. Already solid-state sensors exist which react to temperature, pressure, and the concentration of certain simple molecules. Dedicated microelectronic circuits, or indeed microprocessors, provide the ability to convert a local signal into a suitable form for more central control. In this way, the key economic and technical barriers to full process control in continuous process industries have been removed, in much the same way as for the discrete engineering industries.

A further economic incentive towards automation of continuous process operations is provided by the dramatically increased potential to react appropriately to a much wider variability in raw inputs. Thus, the need to blend slightly different feedstocks in a petrochemical or chemical plant is reduced; the ability to mix together slightly varying waste materials for recovery is increased. Indeed, the situation for continuous processes is exactly analogous to the enhanced ability of discrete engineering processes such as assembly. Sub-assemblies of wider tolerances can be so chosen as to give final assemblies of the same previous tolerance, or even better. These two examples bring out a feature of all the new technologies associated with microelectronics: the ability to bring reasoning power, so-called "intelligence", to the workplace.

Other process application areas

The primary industries, and mining in particular, are already benefiting from increased automation in mechanical handling and in refinement of the extracted raw material. Processing at the mining location has now become feasible, for many metallic ore industries, and particularly interesting are the large-scale experiments in burning coal in-situ.

Deposits of ore and minerals are frequently located in areas not well suited to the use of human labour, a factor which had formerly played a large part in the discounting of known deposits as not economically exploitable. Automation overcomes this limitation, and thus is substantially increasing the number and extent of viable reserve deposits.

Another often overlooked development in this area is the increasing capability of remote sensing devices to locate geological deposits of economic or strategic importance. The whole of the electromagnetic spectrum can now be harnessed to reflect the content of near-surface deposits, and the combination of ground-truth, aerial and space-based survey techniques is being rapidly extended.

Such surveillance techniques have a major part to play, too, in flood warning, earthquake prediction, weather prediction, crop prediction and other spatial interests. The previous high cost of gathering and collating the vast bank of data involved is rapidly being discounted, and cross-frontier international co-operation is developing to produce an international institutional framework for the full exploitation of this newly emerging power.

In agriculture itself, enormous benefits can be expected from the application of improved crop spraying procedures which can be more locally controlled by using microelectronics. Techniques of planting, too, are in a state of change. It is now possible for each seedling to be accompanied by a nutrient dose tailored to the plant's own requirements and to conditions in its immediate vicinity, as measured by electronic sensors and calculated at the time of insertion.

Equally staggering are the rapid developments of new strains of flora and breeds of fauna, made possible by the enhanced computing power now available to the geneticists. Also of great importance, as it spreads throughout the world, will be techniques for the better protection of the growing crop, particularly in greenhouse control and in irrigation control, and for protection after harvesting and whilst in storage. The high level of wastage for many mass crops such as cereals should be reduced, as the new control techniques made feasible by microelectronics come into more general use.

Of course, many of the microelectronic applications in agriculture are expensive, and they require an infrastructure of support far exceeding that of the current "green revolution". But the ability to produce high value crops out of season, or the necessity to bring previously non-available acreage into cultivation, will produce need-imperatives which will ensure quite wide application in the advanced economies. For the main crops such as rice, wheat or corn, increased yields or bountiful harvests will inevitably produce instabilities in demand and supply, and focus attention upon questions of maldistribution and access to purchasing power.

Effects upon the Infrastructure

In reviewing the effect of microelectronics upon the infrastructure of a society, we shall be taking into consideration virtually all of the networks of services which go to create the skeleton of the social body. These include not only the machinery of government and administration, at all levels from the local to the national and indeed international, but also the public services, the financial system, and the major service industries of transport, energy and so on.

One factor which all these disparate sectors have in common is their networked nature: they all consist of a vast enterprise serving an even more vast public, and communication both within the enterprise network itself, and between the enterprise and its public, is a vital factor in their efficient operation. Behind communication lies that which is communicated—information—and it is no accident that the web of technologies of which microelectronics forms the centre is coming to be referred to as information technology.

The flow of information

The convergence of information processing (the role of the computer and the word processor) and communication (increasingly dominated by the telecommunications system, itself a vital part of the social infrastructure) has brought about a revolutionary change in the quality of information flow. More information is available than ever before, it can be processed more efficiently and flexibly than before, and it can be transmitted and acted upon more rapidly. By streamlining the life-blood of the social infrastructure, the flow of information, microelectronics is likely to bring about cumulatively massive changes in the organisation and administration of every society.

Certainly the existing bureaucracy, supported by an efficient information system, might be expected to provide an enhanced public service. Internally, the improved information available to each individual should lead to a reduction in misdirected effort, and to a greater sense of immediacy in operations. Externally, the technical framework is becoming available for fast, effective communication with the general public.

What is feasible technically, however, is a very different question from what is socially and institutionally feasible. For example, much of the application of information technology results in decentralisation and often a flattening of the hierarchy of power. Desirable though this may seem in theory, in practice it seems to ignore two realities. Many people, through instinct, education or attitude do not relish responsibility, whilst others may prefer to retain their positions of power in the status quo!

Again, the merging of a PTT (a postal, telegraphic and telecommunications administration) with other fields of activity may be technically feasible and economically desirable, but so wrapped around with legal, constitutional and administrative problems that progress may be slow. This may well be particularly true when transborder problems of co-operation and collaboration emerge, as is evident in the complicated relationships between various PTTs and satellite-based communications organisations. For example, the revenues derived from submarine-cable based telecommunications are apportioned in a way reflecting earlier history, while the apportionment from a mixed satellite-cable network involves pricing and utilisation policies of considerable complexity.

It is becoming increasingly feasible for government and public service organisations to seek reactions from the public to proposed developments as well as to implemented decisions and their effects. At its most extreme, such a capability might develop into a constitutional system based upon "instant referenda": a prospect which could transform the face of democratic societies. Whether such an ability to provide instant reactions would lead to a society better tailored to its members' wishes, or to a series of dramatic fluctuations in policies, we have yet to discover. The relatively small-scale experiments in opinion-seeking which are presently being pursued in several Western countries may help to provide an answer.

New media

Ideas for "instant referenda" are not new, but microelectronics has brought about the possibility of widespread and cheap realisation. One of the major issues is that often at great expense and over considerable time, checks and balances have been built into the use of mass media and government-licensed broadcasting corporations in many countries. The technical possibilities for new media are starting, but it is not certain that widespread misuse and abuse of the new media can be avoided. The recent UNESCO debates on the role of the press and other media show that different governmental and cultural perceptions of both national and international media genuinely exist. Given the existence of satellite-based television stations shortly to rebroadcast common programmes across multi-country continents, much more will be heard of these issues.

Most media today have a "broadcast" content—for example, radio, television, newspapers, journals; whereas direct interpersonal media such as face-to-face interaction, post and telephone are restricted in scope. The coming of cheap terminals incorporating simple computing ability opens up the way for interactive networks via the telephone system, or indeed radio propagation, whereby one or more people can contact one or more people directly. The rapid acceptance of Citizen's Band radio in the West shows that there is a pent-up demand for such a service, and it is easy to envisage a hard copy message sending service alongside a voice service. It is already becoming common for present home computers, even with their current low volume, to be used in this way, and it may be that such interactive networks will be a tremendous social organising force in the future. Whether such networks will tend to integrate, or to further fragment, social structure is an open question but an important issue. Socially, a parallel can be drawn with the potential opened up in the government world for centralisation and decentralisation. The real issue is that of allowing access to the whole picture of what is going on.

As well as transmitting information, the microelectronically controlled information technology network can be used to transmit money: and this, too, should be of benefit to government organisations. The giro system of money transactions, and in particular social transfer payments, now common in Western Europe will be much facilitated by microelectronic-based computing power, and the potential for flexibility in matching transfer payments to individual or family needs much enhanced.

The same type of system, needless to say, can be employed in the commercial financial sector, and the growth in electronic funds transfer is a major issue facing banks and financial institutions today.

Changes in money-handling patterns have ripple effects of their own right through the social stucture. They affect the retail sector, by revolutionising methods of paying for goods and of obtaining credit. They affect the banks' own delicate balance between their range of operations and their sources of income. And they affect government means of controlling the quantity and flow of money. For these wider reasons, electronic funds transfer is a sensitive subject, and progress towards a large-scale automated system has been relatively slow while the ramifications of its possible effects are being considered.

The load factor

A different aspect of the question of information flow dominates the transport sector. Here efficient operation of the system—whether by land, sea or air, and whether of goods or of people—demands up-to-date information, in order to give needed flexibility to schedules and to marshal transport availability as and when it is needed.

The transport industries as a whole operate on a long life-span for capital investment, and this factor mitigates against their early adoption of microelectronic control and communication systems, except on a piecemeal basis. However, the computerised air passenger booking system provides a good indication of the likely path of development. There is no doubt that the "load factor" dominates the efficiency criteria of any transport industry, particularly the high added-value transport industries of air transport, and unitised shipping (container traffic). Microelectronic control, by offering the opportunity to tailor services to demand, and thus increase averge load factors, offers massive economic advantages.

The load factor problem also dominates the financial calculations of other service industries: electricity, gas, the water supply industry, and the telecommunications system, for instance. These organisations also suffer, to a greater or lesser extent, from the difficulty of coping with peaks in demand, and from the economic drag of spare capacity at periods of low demand.

Microelectronically controlled information systems once again offer rapid feedback upon demand levels, and should facilitate calculation of demand requirements. Even more importantly, they could play a part in reducing the unevenness of the demand schedule. Feedback on user levels could be used to operate a marginal pricing system, considerably more sensitive than any of those currently utilised. Adverse marginal pricing in peak demand periods obviously reduces the demand level: and such a system could serve to reduce the need for planned capital expenditure required only to cope with peak demands.

A particularly good example of microelectronic control capability is to be found in its growing application to the energy production, distribution and use systems. At every level of the national multistage network, improved control procedures are being applied—at the power station, at distributive points within the national grid, throughout district central heating systems, even down to the individual domestic appliances. The gains economically achievable with today's plant are conservatively estimated at some 15 per cent, whilst the developments now made possible in heat-pumps, and even the capture of renewable energy sources such as solar energy, wind and wave power, will make for similar savings.

Education and training

The education and training sector requires rather different consideration, for it will be responding to the changes engendered by the microelectronic revolution, as well as effecting its own internal changes to an even greater extent than the other parts of the social infrastructure. The new generation must be prepared to take their places in a very different world, and one the shape of which, to a large extent, we can only guess at.

Flexibility, once again, is an essential keynote: flexibility both in the education and training system itself, and in the people it trains. Microelectronic applications can help to provide that flexibility. Computer aided instruction, individually guiding the student through a body of material, with the aid of a computer programme that tests comprehension and adapts the pace of learning to the student's ability, is already a tool to be taken seriously. The major limitations to its development are not technical, but financial and practical: in particular, it is expensive and time-consuming to build up an adequate library of applications software.

The use of the computer as a teaching aid may accelerate the process of breaking down the barriers between formal and informal education which can already be perceived. The television today is a major educational medium available as readily (or perhaps even more readily) in the home as in the school or college. Tomorrow the home computer could play a similar, but wider role: providing not only information, but guidance in absorbing and using that information.

While social interaction inside an educational establishment plays a major part in the educational process—and it can be expected to continue to do so—it is nevertheless true that learning at home may become an increasingly important feature of the educational system. If the world fuel crisis continues to intensify, this may provide an added reason for shifting at least part of the learning programme from the school or college to the home. At the university level, indeed, the home is already much utilised as a place of study. One possibility might be a more conscious separation of the social and behavioural training function of education from its purely vocational/educational functions. This is perhaps foreshadowed in developments such as the British Open University, where home study is combined with "summer school" sessions where much of the emphasis is on social interaction.

Social interaction, though, is not the only dimension in which the computer proves to be an inadequate total substitute for the human teacher. Knowledge and information are not the same thing, and to translate mere information into a framework of knowledge is a task which requires more complex abilities than mere straightforward transmission. Thinking is a matter of intuitive leaps as well as of logic, and the computer does not cater for intuition or for fantasy. To use the computer as a substitute for education of one human being by another may open up the prospect of a world is which the ability to use imagination is devalued. It is more desirable to see the computer as a tool for the teacher, than as a substitute for him.

As well as providing an educational location, the home may also become increasingly important as a location for work and leisure activities. The spread of information technology into the home—through the home computer and the telecommunications net-work—increases the range of activities which can be pursued at home. Executives can access their office mail through their home terminal; housewives order goods via view-data systems; the entire family obtain access to a wide variety of entertainment in every conceivable medium.

Such developments, yet again, depend not upon further technological breakthroughs but upon the application of technology which is in existence today, and upon the process of choice as to how that technology is to be applied. And if they do take place, they in their turn will demand further adjustments in the social infrastructure. The impact of microelectronic technology and its applications cannot be considered in isolation: every choice, every change, leads to other choices and other changes.

Exploding Opportunities: Microelectronics in Research and Development

One recurrent theme in this chapter so far has been that very few of the developments we have mentioned are dependent upon further technical advances. Microelectronic technology is already far advanced, and we have a long way to go before we have explored all the application possibilities currently open to us. A minority of potential applications will demand still further levels of sophistication from their microelectronic control systems, but they are only a minority. What we already technically know how to do will be more than sufficient to totally transform society.

Research and development in the microelectronic field is continuing, naturally, and the graph for projected increases in processing capability continues to rise steeply throughout the remainder of the twentieth century. But of wider interest is the effect which the level of microelectronic technology currently available will have upon research and development in quite different areas.

Research is inevitably something of a hit-or-miss affair, and control of the direction of research, and evaluation of results, are both information-intensive activities. Computing power can be harnessed to enhance both of these aspects of the research and development function. The developing global information system can do much to ensure that researchers are made aware of parallel or closely related endeavours, and that (where political or economic barriers do not intervene) information is exchanged fully and rapidly among the international research community. The computer's sheer number-crunching abilities can fish out significances from apparently meaningless research results, and channel the direction of future research in potentially profitable directions.

And microelectronics will, of course, have end-user applications in many scientific fields being developed today: monitoring and controlling complex or sensitive operations, and thus making an entire new range of operational techniques practicable.

Typical of many of the new processes and products to be expected, as the new wave of efficient and almost universal computing power makes itself felt in the research and development laboratories, is the potential of biological processing in industrial chemistry. The advantage of such biological processing—for example, enzyme catalysis or biological leaching and concentration—is that they involve processes carried out under normal ambient temperature and pressure. The high temperature and pressure processes developed after the Second World War are increasingly vulnerable to shifts in energy prices, and often involve more extreme environmental hazard. Additionally, biological leaching offers a route to the utilisation, with high environmental gain, of the vast wastes and residues of industrialisation.

Already a considerable amount of animal feedstuff is being produced by bacterial fermentation of crude feedstock to single-cell protein. Although the feedstuff to animal to human protein route is inefficient, at least this development will release land for growing food for direct consumption by humans. All these processes were unthinkable without the control afforded by microelectronic development.

In the energy field, advances in heat storage—for instance, the use of very high heat-capacity materials such as the metallic hydrides—offer higher efficiencies in energy usage, as do the now economically controllable usages of heat pumps to upgrade energy gradients. The renewable energy sources of wind, sun and tide are now more harnessable and utilisable, given the flexibility of both local and network-based microelectronic controls which can compensate for their sometimes unpredictable variability.

In this kind of way, the whole research and development system can be accelerated in its output of usable devices and processes. Virtually unlimited computing power is now nearly to hand, and a cornucopia of new developments can be expected, dwarfing the already staggering level of output.

Rates of Diffusion

Two factors make it particularly difficult to predict the rate at which these actual and potential applications of microelectronic technology will spread throughout world economies. The first is, quite simply, that the question is not a technical one but an economic one. It is not the technical feasibility of most of the applications we have described, but the economic capability to take advantage of them, which is in question. And as such, the question of the rate of diffusion of this technology is inextricably bound up with the question of the general economic future—a vast and impenetrable subject.

Secondly, the industry is still an immature one. Some sectors can be considered to be reaching their stage of maturity, but in many that phase still lies in the future. The rates of growth we have already witnessed, then, cannot be extrapolated into the indefinite future. They do, however, provide a general indicator of the current pattern of development of the applications industry.

Current rates of growth

The following table shows the current (January 1981) rate of growth of microelectronic equipment being delivered to various application fields. In total value this equipment is running at several scores of billions of dollars in value, but it must be remembered that the total value of incorporating equipment is up to ten times higher on average.

	Per cent per annum
Data processing systems	8
storage	12
peripherals	18
Office copying	25
Word processing	30
Office, other	15
Communications: telecommunications	24
radio/television	14
data communications	11
Industrial controls	18
Test and measure	14
Automobile	22
Medical equipment	15
Other controls	17
Consumer audio	4
Home appliances	6
Personal devices	8
Video recorders	9
Games, etc.	14
Television	6
Defence	28

The basic microelectronics manufacturing capability, in producing microprocessors and relatively small (up to 16,000 bits) memory chips, matured in about 1978. Interestingly, the first simple microprocessor was especially designed by Intel for the first Japanese calculator. Previous concepts of mass markets had been confined to use in direct substitution for the computer, and computer-related fields. Of course the rapid take-off of the calculator chip was followed by a similar expansion in digital watch circuitry, and another "free-standing" application, that of television games, soon followed. It has been estimated that well over a billion pocket calculators have been produced in the last six years, and some seven hundred million digital watches. Approximately three hundred thousand television games have followed. All these markets show the rapid price fall which such elementary products will exhibit as price competition takes place.

The incorporation of microelectronic capability into other consumer goods such as washing machines, cookers or into capital goods such as typewriters, word processors or machine tools has proceeded apace. Where substitution or improvement could take place without major redesign, innovation has only been partly restricted by the product design and life cycle. A major feature has been the appearance of new entrants in product areas. These new entrants are often based on adjacent product fields where electronics and electronic capability was already more advanced.

Major factors affecting the diffusion rate

One major barrier which has been claimed to exist, and which presumably would have slowed down the rate of diffusion, is a shortage of electronic technicians and a shortage of training in the new field of microelectronics. The shortage, however, is somewhat exaggerated, since there is a relatively high mobility of the often young technicians which gives rise to an active labour market.

In the case of more major capital goods, for example in the telecommunications field, two factors have disguised the high latent rate of change. The demand of such fields is for highly specialised chip circuitry, and only now is the demand sufficiently large to warrant full-scale production of these special purpose products, as opposed to the more general purpose microprocessors.

A second and even more crucial factor is a need to redesign the rest of the control or instrumentation system to match the electronic capability—a factor often compounded by the need to achieve industry-wide, even sometimes international, agreement upon standards. In early 1981 it is, however, true that the telecommunications industry and associated equipment is the fastest expanding user of microelectronic output, at a level rapidly approaching that of the computing industry based on data processing.

The opportunities opened up by microelectronic capability have been a strong factor in the pressure for deregulation of the traditionally publicly controlled postal, telephone and telegraph monopolies. The blurring of the boundaries between public telecommunications, broadcasting, Citizen's Band radio, distribution of electronic images within and between offices and organisations, and computing and data processing, has meant that many small firms have seized the opportunity to fill market gaps which have appeared amongst the products of traditional suppliers. Many such firms buy in components and housing, and can bring new products to the market more quickly than larger firms.

Such new firms are often founded by entrepreneurs who have left larger organisations, a mechanism which reflects the origins of the semiconductor industry itself. As a measure of the degree of innovativeness, some thirty to forty new electronic firms are registered daily in Western Europe and the United States alone. Current forecasts suggest that only half will survive longer than three years or so, but one or two will go on to become medium-size employers.

Conclusion

This necessarily short survey of microelectronics in application may give the picture of a bewildering diversity with little pattern. This would be misleading. There is a common feature to all these applications. Microelectronics essentially facilitates communications, instrumentation and automation—every application shows what has sometimes been called the coming of the Age of Control.

The pattern of future applications will follow this imperative, as in turn every productive activity comes under scrutiny and the question is asked, how can it be controlled to make it more efficient? To the electronics medium, the content can represent data, voice, telex, text or image; to the world-wide communications network distance and time are shrunk away. Expert knowledge on any subject can be delivered in response to any problem, and if the pattern is sufficiently repetitive, an automatic machine put in place to respond.

None of the developments we have described are less than technically probable. Most are technically possible today, and many application products and processes are already in existence. The speed of diffusion of these applications depends, primarily, not upon technical factors, but upon economic and social ones. It is not a question of what we can achieve from a technical standpoint. It is a question of what we want, and how our new technological capabilities can be harnessed to fulfil those wants. The major constraints will be institutional, and the battle therefore will be between the vision of those who can see how wants can be fulfilled, and the vision, or lack of it, of those peopling our institutions.

Inevitably there is a negative side to some of the applications now feasible. Development along some routes, if not all, would lead to an increase in sources of tension; to a widening of economic gaps, and to a number of foreseeable social and political problems. But at the same time, there is, we believe, a very positive side to the outline we have sketched. Microelectronic technology has so much to contribute to the alleviation of problems facing the world today. Applied in these positive directions, it could help to solve the world's energy crisis; enhance its reserves of food and of raw materials; anticipate and prepare for natural disasters; deepen our sources of knowledge, and enhance our level of communication with one another.

Above all, microelectronics is a technology we already have: and it is being launched on a world we already have, with all its strengths and weaknesses. It is up to us to ensure that we use this technology to change our world for the better.

4

The Impact on the Enterprise

BRUNO LAMBORGHINI

Radical Changes in Enterprises

Consequences for the whole enterprise structure

Microelectronics has already caused radical changes in the organisation, structures and strategies of enterprises, especially in manufacturing.

In the future, the impact of microelectronics will spread across the entire structure of an enterprise from manufacturing to administration, from planning to marketing. Microelectronics will have radical consequences not only for companies in specific industrial sectors more directly involved with technology, such as components and electronic products, but for all kinds of enterprises or economic activity or organised structures.

Since the 1970s, enterprises have found themselves operating in a climate of general economic instability, frequent and unpredictable cyclical fluctuations, generalised slow-down in growth, reduced productivity, rising costs in production and particularly in energy products, and as a result, low profits and inadequate cash-flow for expansionary investments. Together these factors lead to a general slow-down in production which means a lack of new jobs or even considerable cuts in employment. High inflation discourages enterprises from making investments, particularly expansionary investments, whilst resources are concentrated on labour-saving and energy-saving investments. It does not seem likely that this situation will change radically during the 1980s while the structural conditions causing it persist.

119

The labour-saving effect

The introduction of microelectronics is part of the picture and tends to aggravate problems and change preoccupations, especially in relation to labour-saving and labour-substituting effects. At the macroeconomic level it is difficult to establish which direct labour-saving effects are really of technological origin as opposed to those due to general economic factors which up to now have been considered the greatest cause of unemployment.

At the level of the individual enterprise, however, the direct consequences of the introduction of microelectronic technology are more evident, since it particularly leads to the substitution of mechanical products with a high labour content by microelectronic components with a decreasing labour content, or to the introduction of automated systems and robots into manufacturing processes—both reducing manpower.

At the enterprise level, microelectronics does not simply open up previously unheard-of opportunities to improve efficiency and productivity through the spread of automation and control of manufacturing, managerial and decision-making activities, but it also enables the enterprise to readjust effectively to the new conditions imposed by profound economic and social changes.

General readjustment

This readjustment can be achieved through greater flexibility in adapting production processes to shorter and less predictable product life-cycles, through significant reductions in energy consumption, through the development of new products and markets, through a more rational use of information in decision-making and planning.

This last aspect is the most truly innovative consequence of microelectronics: the capacity to "produce", collect and diffuse enormous quantities of high-quality information at minimal cost. The use of microelectronics in manufacturing, administration and planning activities enables growing quantities of information to be placed at the disposal of enterprises at a decreasing cost and with increasing ease of access.

A new product: information

In manufacturing, the "new" information created by the introduction of microelectronics can modify old production functions and determine new cost/benefit ratios.

The innovation brought by the new technology does not correspond to a simple process of substitution of human labour with machines: unlike previous automation processes, the real innovation of microelectronics is the quantity and quality of the "new" information which is the "principal" product of microelectronic applications.

Information becomes the predominant, even if non-material, factor in production functions and decision-making within enterprises. Only if it is appropriately used, can the flow of information enable enterprises to adjust effectively and to achieve positive socioeconomic results. That is, technology provides here an opportunity. If the opportunity is not taken, microelectronics may in the future have catastrophic consequences for many enterprises, decreasing their competitiveness and increasing economic and employment crises.

The structure of the enterprises involved with microelectronics is profoundly modified as manufacturing, organisational and marketing adjustments become necessary. The degree of flexibility required is higher than for any previous technology. Microelectronics exposes any inherent weakness in an enterprise's structure, its managerial capabilities, its market positions. This occurs, in different but not dissimilar ways, both in companies which incorporate microchips in their products and in those which use microchips and microelectronic equipment in production, administration, management and service activities. In some cases microelectronics requires the complete restructuring of an enterprise (for example when the old mechanical technologies are totally eliminated, as in digital watches) and in others it necessitates total manpower retraining or plant modifications (for example from a large assembly unit to small specialised workshops or from a highly fragmented labour organisation to full robotisation).

The diffusion of microelectronics is out of control

The rate and manner of the diffusion of microelectronics in products, in production and in administrative processes are basically beyond the control of single enterprises. Even very large enterprises operating in sectors where a strong oligopoly exists, such as large computers, are also finding it difficult to effectively control the rate of innovation and product life-cycle, due to the evolution of microelectronics.

In the case of electronic computers, it is in the interests of all manufacturers with a large base of installed rented machines to control and slow down for as long as possible the introduction of new systems with radically innovative technological features, in order to keep the machines on rent installed for many years and thus ensure a high return on investments.

The dramatic developments in microelectronics at the end of the 1970s have enabled new enterprises to enter the computer market with systems which are highly competitive in terms of price and performance and totally compatible with the software of the leading manufacturers (Plug Compatible Manufacturers or PCMs). The pressure exerted by the PCMs has led to an acceleration of the rhythm of innovation among the major manufacturers who would otherwise have delayed the introduction of new products.

The introduction of the new IBM 4300 computers, a revolutionary product in terms of price and performance, was brought forward to the beginning of 1979 in a countermove against the PCMs. In this case the developments directly generated by microelectronics outside the companies established in the specific market have set off a deep-rooted process of innovation in the entire computer market, which is beyond the control of the major enterprises. Endogenous and exogenous developments therefore tend to accelerate the innovatory process.

The pervasiveness of microelectronics

This phenomenon seems to apply exclusively to microelectronic technologies. In fact, the special features of microelectronics are its pervasiveness and the rapidity and relative ease with which its use spreads.

The rapid diffusion of microelectronic applications is stimulated and propelled by the worldwide presence of a large number of semiconductor manufacturers, among whom competitiveness is necessarily high because of the extremely reduced life-cycle of their products.This factor also causes manufacturers to implement competitive economies of scale in production as quickly as possible.

The relative ease of the spread of technological know-how, which is more closely related to the capabilities of individuals or of research groups than to patents, is demonstrated by the development of the world's leading microcircuit production centre, Silicon Valley, in California. In the space of ten years, what was originally a Fairchild base has been developed, partly because during the 1970s this type of production required only a modest amount of fixed capital, into a dense network of small and medium companies which tend to accelerate the rate of technological development.

This exceptional growth causes difficulties both in miniaturisation (for example, the production problems and the low returns of 64-K RAM chips) and, above all, in finance, due to the rising equipment costs in manufacturing. However, whilst these difficulties may have negative effects on some companies in the sector, they do not seem to slow down technological development. Sometimes a slowdown in the rate of technological growth is caused by large users' reactions, as is the case of PTT (postal, telegraphic and telecommunications) authorities and the procurement of telecommunications equipment; in large electromechanical telephonic switching systems a rapid reduction of the forecast period of depreciation creates difficulties and untenable costs. In this instance, the spread of microelectronic technologies (which means the complete restructuring of the switching and transmission systems) is regulated by PTT's replacement plans and not by those of the manufacturers.

Software is a bottleneck

But there is a much more significant factor slowing down the spread of microelectronics which is outside the companies' capacities for innovation and increases difficulties. This factor is software production, particularly application software, which is growing less quickly than the production of microelectronic hardware.

A computer without a programme is like a car without fuel. But hardware technologies are proceeding at much faster rates than programme preparation technologies. The problems caused by the manpower shortage in software are considerable. In 1980 demand for computer programmers in the United States exceeded supply by at least 50,000 people, and this gap is going to increase enormously. Software is therefore an extremely serious bottleneck in the development of microelectronic applications.

The problem of how enterprises can adapt positively to the challenge of microelectronics can be resolved with the minimum of social upheaval and the maximum utilisation of technological potential only if enterprises can adequately handle the radical conversion of their human resources, both workers and management, in a climate of consensus. This requires heavy investments in training and retraining and permanent education related to the deep-rooted and continual changes which microelectronics will bring for a long time to come. The problem of making a special commitment to training and of actively readapting company personnel can be handled within the enterprise only to a limited extent; it must be solved above all through public training policies.

The Impact of Public Policies on the Enterprises

The increasing influence of public policies in general

The role of governments in planning the rate of technical innovation through regulatory, fiscal and credit policies is already extensive and is likely to become even more so in the future, as governments attempt to control the direction of technical change.

The development and spread of microelectronic products and applications within enterprises is everywhere increasingly influenced by public policies. These consist of direct and indirect measures, regulations and incentives which actively affect the policies and activities of enterprises.

Government intervention is not only concerned with the growth rate of technological innovation (support policies for research), but also, and increasingly, with the applications of microelectronics and their effects on the socioeconomic situation.

In some countries complex public policies using various tools have been developed, and these are becoming an increasingly important factor in decision making at the enterprise level.

The case of Japan

The creation of an adequate industrial structure in the field of microelectronic components is an objective common to Japan and all the major European countries. In 1976 Japan set up a VLSI (very large scale integration) research project, which is jointly conducted by major computer manufacturers in co-operation with the Nippon Telegraph and Telephone Corporation, and with financial backing from MITI (Ministry of International Trade and Industry).

The participants are researching the possibility of processing silicon slices with minimum line widths between 0.1 to 0.5 microns. This will enable the Japanese industry to accelerate technological development at a startling rate from now until the year 2000, and to increase its own competitiveness on international markets, particularly in relation to the U.S. industry.

The case of the United States and Europe

Co-operation among enterprises is also considered fundamental to the microelectronic programmes of the British, French and West German governments, which provide incentives for joint development programmes among their national enterprises, and also with U.S. companies in order to accelerate technology exchanges with the most advanced country in this sector.

The U.S. Government supports its microelectronics industry through an extensive procurement policy for space exploration and military purposes, and through direct support by the Department of Defense for the development of VHSI (very high speed integrated) circuits.

In all these countries government intervention in microelectronics is increasingly directed towards policies concerning demand and applications.

These policies concern the large application sectors, such as data processing and telecommunications, where in general specific industrial conversion programmes are developed to adapt to the impact of the new microelectronic technologies. They also single out certain areas. For example in France, CODIS (Comité d'Orientation pour le Développement des Industries Stratégiques) has concluded public development contracts for selected areas (particularly office automation, robotics and consumer electronics); and in the United Kingdom the Advisory Council for Applied Research and Development has proposed candidates for SRC (Science Research Council) support, including: robotics, measurement and instrumentation, microprocessor work benches/teaching aids and man/machine interface. Furthermore, special projects are developed to help companies study microelectronic applications, with training courses for small firms. Public policies and programmes pay in schools and universities increasing attention to microelectronics. However, the activities of the various bodies seem to be poorly co-ordinated. The object of industrial intervention policies is to introduce new products on the international market ahead of other countries in order to win better positions in international trade.

EEC action

The action proposed by the European Economic Community in the field of microelectronics intends to overcome the piecemeal intervention of the various member countries by developing co-ordinated policies to improve the competitiveness of the European industry against the current leadership of U.S. and Japanese industries. The EEC Commission stresses the danger of dependence on outside suppliers, and criticises the national initiatives taken by individual countries as nothing more than a palliative. However, the European Community seems unable to impose on its members a unified approach to intervention policies. It tries to convince enterprises to co-operate with one another, to exchange technical information, to undertake joint ventures. The EEC considers the European market to be a real chance for the European microprocessor industry; thus it is making efforts to standardise and making recommendations to open up public purchasing policies even in such extremely restricted areas as telecommunications.

Many countries have taken steps to assist a better understanding of how microelectronics will affect enterprises and employment in the future, and to make society aware of the problems caused by microelectronics (e.g. the Nora-Minc Report on the informatisation of society for the French Government, and the 1980 Prognos Report for the German Federal Republic Government on the impact of technology on employment).

Unco-ordinated approaches

The policies developed by the various countries thus appear fragmentary, unco-ordinated and inadequate in the face of such a complex problem as the impact of microelectronics on the industrial system and society. They tend to give support, often of a *laissez-faire* nature, to enterprises mainly in order to increase the competitiveness of the given national industrial system. They rarely make an overall evaluation of the socioeconomic consequences which this action may have on the national and international social and industrial system. They often impose development-blocking restrictions and regulations on enterprises, without providing medium- or long-term guidelines. There is no comprehensive blueprint for the type of society to which the guided development of microelectronics and other emerging technologies should lead.

On the other hand, financial support tends to be provided for the survival of "lame ducks" which have been unable to adapt to technological change, either because of their own inability or because of external restrictions. There is a real danger in some countries that by slowing down the conversion process, public intervention will aggravate and reinforce the crisis which microelectronics can generate in many enterprises.

But basically there seems to be a lack of adequate planning as regards education to provide the skills needed for the adaptation of personnel in industry and services: in the next few years this could seriously limit the enterprises' ability to adapt to the new conditions created by microelectronic technologies.

Microelectronics Incorporated in Products

The spread of applications

The spread of microelectronic applications increases industrial activities which incorporate microelectronics in their products: it is expected that a considerable number of industrial sectors using mechanical technologies will convert to electronics in the next few years. Growth rates in the different sectors using electronic components are significantly higher than the average rates in the non-electronic manufacturing industry.

Furthermore, microelectronics modifies the boundaries between the different industrial sectors, by encouraging the convergence of different industrial areas and different enterprises. The most obvious example is the convergence of the data processing industry and the telecommunications equipment industry. This convergence, which is a direct consequence of the evolution of microelectronics, is actively developed by large multinationals and by governments through acquisition and merger policies. The "telematics" industry of the future may lead to a high concentration of industrial and service activities in the hands of a few extremely large companies operating on a world scale.

On the whole, the industries extensively incorporating microelectronic components in their products are expected to become the largest and the most rapidly expanding ones in the world in the 1980s, coming ahead of the two most influential industrial sectors in economic growth during the sixties and the seventies: the petrochemical industry and the automobile industry. Forecasts predict that total revenues in the world's electronics industry will jump from $250 billion in 1980 to nearly $800 billion (at 1979 prices) in 1990.

Changes in the manufacturing cycle

The most well-known and most discussed phenomenon which is directly caused by the development of microelectronics is the sharp reduction of work phases in the production cycle (see Fig. 1).

Moreover, the impact of microelectronics on these enterprises is an

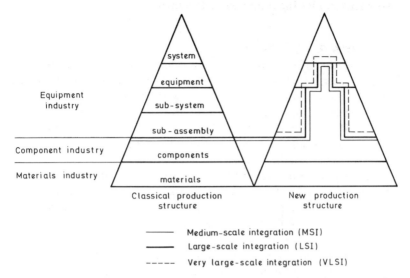

Fig. 1. Increasing integration in electronic components sharply reduces production
work phases in the equipment industry.

even more extensive phenomenon affecting all aspects of company
life: from research and development to marketing, from planning to
administration, from product policy to personnel policy.

Particular attention should be paid to the deep-rooted changes
which microelectronics causes in the manufacturing cycle. In general
terms, this cycle can be summarised as a succession of transformation
phases in which parts and components are made from raw materials,
and sub-systems and end-products from assembled parts.

The spread of microelectronics has radically altered this process.
Developments in LSI (large-scale integration) components, and the
prospect of even greater developments in VLSI (very-large-scale integration) components, make it possible to build a growing number of
previously separate functions and already assembled parts and
systems into the components. As component integration increases, the
number of phases of the manufacturing cycle is reduced (for example,
see the next section). Growing component integration, which is accompanied by a marked reduction of costs for parts and elementary

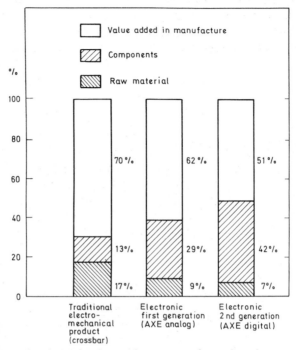

Fig. 2. How microelectronics changes the structure of manufacturing costs. Source: *Financial Times,* 25 September 1980.

functions, results in a shift of manufacturing activities and of value added from end-product enterprises to microelectronic components companies. (Moreover, as components become increasingly integrated and miniaturised, the manpower requirements of semiconductor manufacturers are also decreasing, since more highly automated production processes with fewer assembly stages are needed.) Since the speed of innovation and economies of scale have led in general to a separation between the production of electronic components and the production of systems and end-products, there is sometimes a rapid decrease in the manufacturing activities of end-product companies and a drop in the level of production integration.

In telecommunications where value added for traditional electromechanical switches is 70 per cent of the manufacturing costs, there has been a reduction to 62 per cent with the first generation electronic switches and to 51 per cent with the second generation switches. The corresponding value of external components and parts in the manufacturing costs has increased from 13 per cent (mechanical exchange) to 29 per cent (first electronic generation) and to 42 per cent (second electronic generation) (see Fig. 2).

Substitution of mechanical parts

The substitution of parts and sub-assemblies with microelectronic components is particularly evident where there is a conversion from mechanical to electronic products; the most obvious examples include office machines, telephone switching equipment, watches, instrumentation, etc. However, it can also be found, and is significant, in electronic products which are affected by successive developments in microelectronics: e.g. computers, colour television sets, home applicances, etc.

There are numerous examples of the reductions in value added and in the production phases, as a result of this substitution process. For instance, an electronic watch requires the production and assembly of only five components compared with some 1000 assembly operations for a mechanical watch. Or, the production of an electromechanical teleprinter requires more than 75 hours, whilst an electronic teleprinter needed only 17.7 hours in 1980 (one microprocessor can replace up to 936 mechanical parts). An electronic taximeter requires 3.7 hours compared to 11.7 working hours for an electromechanical one. Over 9 hours are needed to assemble a mechanical calculator, whereas a printing electronic calculator requires less than 1 hour. An electric accounting machine takes 33 hours to produce, compared to less than 13 hours for an electronic small business system. An electromechanical typewriter requires 20 working hours whilst electronic models in 1980 took only 10 hours to produce and in the near future it will be 6-7 hours at most. In the manufacture of sewing machines one microprocessor replaces 350 mechanical parts. And so on.

This radical transformation also takes place when electronic components are replaced by more highly integrated components: take for example the developments which have occurred in mainframes, minicomputers and personal computers. The end of this process of transformation and development is still a long way off, as increasingly miniaturised and low-cost generations of technology continue to appear one after the other. For this reason companies which use electronic components have to take into account the continuing impact of a technology which is far from being stable.

Heavy cuts in manpower

During the 1970s and in the early 1980s numerous enterprises have had to make heavy cuts in their labour force, particularly in manufacturing activities where certain jobs have been reduced, especially in metalworking (punching, boring, welding) and component assembly. Checking and testing functions have also been radically modified by microelectronics. Below are some examples for this trend.

The British information processing industry as a whole registered a 20 per cent reduction in employment between 1970 and 1977 despite substantial increases in production volumes.

In 1980 a large computer company formulated plans to close one of its factories and dismiss 2500 people working in metal fabrication. Despite an annual 20 per cent growth in production volume, this redundancy still arose because of the advances in microelectronics.

If one takes the example of four American, three West German and one Italian mixed companies (that is with mechanical and electronic products), there was a reduction in employment close to 20 per cent between 1969 and 1978.

Within these overall employment figures, there was a strong reduction in the number of production personnel: the percentage within the total personnel fell from 45 per cent to 25—30 per cent between 1970 and 1980.

The percentage of blue-collar workers tends to decrease dramatically: for example, in the Italian information processing industry the percentage of blue-collar workers in total personnel decreased from 53 per cent to 38 per cent between 1970 and 1977.

A major U.S. telecommunications manufacturer reduced its labour force from 39,000 in 1970 to around 17,000 ten years later.

In the United Kingdom employment in the production of telephones dropped from 88,000 in 1974 to 65,000 in 1978.

A major Swedish telecommunications manufacturer reduced its work force from 15,300 in 1975 to 10,300 in 1978 (in manufacturing activities from 8000 to 3000). By the time the changeover to electronic equipment is completed, the company expects to have halved its labour force, despite strong increases in production.

In the watch-making industry, employment in Switzerland fell by 46,000 in the 1970s, while in the Federal Republic of Germany employment is estimated to have fallen by 40 per cent during the same period, due to the direct and indirect effects of technological change.

The impact of microelectronics has brought about substantial changes in the structure of enterprises working in the computer sector and even more in those producing office machinery; in the latter case this is due to the radical replacement of highly labour-intensive mechanical or electromechanical parts with increasingly integrated components produced by automation, which involve very low labour content and are manufactured by companies (the components manufacturers) other than those producing the machines.

Changes in the information processing sector

Besides causing the convergence of the two sectors of computers and office machinery into a single sector—information processing—the new technology has greatly accentuated competitiveness, enabling newcomers to enter the market, modifying old market positions and creating opportunities for new applications in both the professional and the private sectors of demand.

Within information processing three different types of industrial structure can be currently identified:

(a) companies using a mixture of technologies ranging from the mechanical production sectors (office machines) to electronic technologies;
(b) companies which have always operated mainly in the field of electronic technologies;
(c) companies with a high degree of application specialisation or involved mainly in software and service activities.

The development of microelectronics has had a different impact on these different types.

In fact, the replacement of mechanical parts with microelectronic circuits causes changes in value added. These changes are clearly greater than those in the companies operating with electronic technologies of the previous generation. However, it should not be forgotten that the microelectronic revolution has also affected production which was already using "old" electronic components, through the elimination of assembly stages and, above all, by making conditions more favourable for "buying" rather than "making" components or finished products.

At the moment even the largest companies make use of external supplies of components, parts and equipment for reasons attributable to cost factors or the speed of innovation.

Rear and forward integration

The big end-product manufacturers are attempting a rear integration strategy by expanding into the microcomponents area, mainly through acquisitions of semiconductor companies, in order not only to recover some of the value added they have lost, but also to control the more strategically important component supply sources.

Vice versa a forward integration strategy is being carried out by microcomponent manufacturers, who wish to penetrate into the equipment and end-product market (mainly in data processing, toys and entertainment).

Given this forward integration, an increasing concentration of power among component manufacturers and in particular a migration of value added to U.S. component sources seems inevitable. This may create dangerous bottlenecks and shortages of components, the first signs of which were already apparent in 1979.

End-product manufacturers, especially the larger companies, are aware of this danger and are trying to secure control of component supplies by creating in-house CAD (computer aided design) centres for circuit mask design or by acquiring semiconductor companies.

It is foreseeable that in the next few years large manufacturers using microelectronic components will find it hard to compete and survive without internally producing such components. Given the high and rising costs of investment to set up an efficient CAD centre or to acquire new equipment for the manufacture of components (an electron-beam microlithography system may cost as much as $2 million, a testing station more than $500,000), it is obvious that the recovery of vertical integration will be achieved only by the strongest enterprises and that such recovery will lead to a significant degree of industrial concentration.

Microelectronics tends to transfer the specificity of individual products to the applications for which the products are intended.

In the area of instrumentation, new kinds of equipment are being developed, defined according to the software used. The same device is in fact able to perform different tasks, e.g. it can function as a spectrum analyser and as a sampling oscillograph.

In information processing, a single machine may act as a typewriter, teleprinter, data terminal and computer, according to the programmes used.

The "black boxes"

It is possible that in the future a large quantity of microprocessors, memories, line controls and peripheral controllers will be incorporated in a single standard "black box". It will be possible to produce millions of these "black boxes" at minimum costs. The application will be assigned according to the specific software "loaded" in the memory of the "black box", which will then be connected to the equipment designed to perform the specific applications which it controls.

This standardisation of microelectronic hardware will undoubtedly affect industrial structures. On the one hand, the production of "black boxes" will be concentrated in the hands of a few manufacturers who can reach exceptional economies of scale: for example, the intention of Japanese industry to reach highly competitive levels in producing standard hardware is well documented. On the other hand, it compels companies in the various application sectors to offer specialised products which, though using standard microelectronic hardware (purchased from external suppliers or manufactured internally in very large companies using highly automated production processes), will have to be specified—if they are intended to be competitive—in terms of the application software and the services which companies can supply to the final user.

Thus, companies may be radically transformed as manufacturing activities decrease and the production of application software and customer services develop. Such a transformation would require an extensive change in the companies' managerial and organisational structures. The disappearance of many companies is therefore inevitable.

Company manpower would also be significantly affected, with a growing reduction of blue-collar workers and an increase of specialists and skilled technical staff.

Although microelectronics causes increasing hardware standardisation, it may enable the application field to be considerably expanded. Application specialisation may be extended through the use of software programmes which are able to perform personalised tasks for specific users.

Standardisation versus "customisation"

The extraordinary ability of microelectronics to make such highly custom-tailored applications available can only be exploited if the development costs of personalised programmes can be limited, that is, if developments in languages and operating software enable these applications to be developed free or at low cost.

At the moment cost considerations lead companies (with the exception of very small companies) to offer standard products, consisting of standardised software packages as well as standard hardware. The internationalisation of markets, a typical consequence of microelectronics, encourages standardisation.

However, the opportunities opened up by microelectronics to achieve effective product "customisation" may be exploited in the future. It is difficult to forecast whether technology alone will be able to direct industrial and application developments towards personalised products rather than, as up to now, towards growing standardisation. In the first case there would be a decrease in mass production and small companies would spread, characterised by more flexible automation and perhaps artisan-type activity; in the second, besides more highly concentrated and more rigidly automated structures, there would certainly be considerable savings in labour.

The resolution of this dilemma of standardisation-personalisation (or "de-massification") may have an even more significant impact on the future of society. Basically, although the most obvious reaction of enterprises in the field of electronics to the impact of microelectronics seems to be the reduction of redundant work force and the formulation of programmes to save labour and reduce costs, the transformations taking place run much deeper.

Flexibility in production

It should be emphasised that the rate of microelectronic innovation (stimulated not so much by the user companies as by the competitiveness among components manufacturers) has reduced the average life of products (from ten years for mechanical accounting machines or five for computers of the 1960s to two years for current minicomputers, from twenty years for mechanical telecommunication switches to five or seven years for electronic ones). The phenomenon creates need for considerable flexibility in production, which involves a departure from the traditional assembly lines to more integrated structures. This, in turn, demands greater responsibility in decision-making and planning activities, and also in commercial distribution structures. It also increases the need for greater resources for the research, design and application developments of hardware and software in order to keep up with the market's rhythm of innovation: compared to the research costs of 1—2 per cent of the sales price for mechanical office machines, some 6—10 per cent are needed for the new electronic equipment and systems. In the case of a U.S. manufacturer, development costs for telecommunications switching systems jumped from $40 million for mechanical technology to $500 million for electronic technology fifteen years later.

In the information processing industry, the increasingly low-cost availability of hardware (central units, memories, peripherals) which tends to be a standard product of specialised manufacturers, is accompanied by a growing importance of operating and application software.

From manufacturing to service companies

It seems that all this will lead to a radical transformation of the employment and organisational structures of the companies in the information processing industry. They will rapidly convert from being essentially manufacturing companies (distinguished by highly integrated production) to being mainly concerned with the assembly of components produced on a vast scale by specialised manufacturers or to becoming suppliers of software, applicative solutions, technical assistance, and, in the final analysis, services.

This trend towards a more service-oriented approach is exemplified by the strategies adopted by some of the leading information processing companies in Europe.

It seems useful to analyse here the new strategy adopted by one of these enterprises as regards both the reorganisation of its factory activity and the general reorganisation of company objectives.

An example of radical change

The company in question experienced the impact of microelectronics to a considerable extent during the 1970s, since the proportion of electronic products in its total production rose between 1970 and 1980 from 15 per cent to approximately 85 per cent.

Radical transformations have taken place in assembly work. Where the assembly of traditional mechanical products was sequential in character, consisting of short-stage integrated mechanical linkages, technological conversion has caused this type of production organisation to be completely abandoned. In fact:

(a) The average commercial life-cycle of the products has drastically shortened (from fifteen years for mechanical calculators to two to three years for the electronic ones), and even the same model may undergo modifications while it is in production due to microcircuit developments.

(b) It has become more difficult to predict market requirements, and the need for flexibility in varying and modifying the product mix has grown considerably.

(c) The increase in the number of products (95 in 1965 and over 600 in 1978/80), the shift of production from simple office machines to complex processing systems, and the growing complexity of performance of office machines have created need for flexible structures to reduce both work in progress and the stocks of finished products, in order to be able to respond promptly to market variations.

(d) The necessary improvements in quality (particularly vital in electronic products) are hindered by the fragmentation of jobs, which makes costly checks and "returns" of defective products unavoidable; moreover the electronic product has functional modular features which enable intermediate testing to be carried out during assembly.

After initial experiments, new organisational structures, subsequently named Integrated Assembly Units or "assembly islands", were gradually extended to all electronic products (and also mechanical products). These "islands" consist of a certain number of workers (according to the product, up to a maximum of 20—30 persons) with full responsibility for inspection and repair, whose objective is the production of a functional tested module or a finished marketable product.

In order to utilise overmanning resulting from the previous production system for mechanical products, the company formulated a diversification strategy, creating new initiatives in specific areas where technical skills and the possibility of obtaining profitable results exist. In the early 1970s the internal machine tool production units were formed into an autonomous company where electronic technologies were subsequently adopted: this new initiative rapidly led to the company becoming a leading producer of numerically controlled machine tools. Other internal production units were subsequently disincorporated into autonomous companies in the areas of electric motors, electronic control equipment, mechanical components, and other similar initiatives are planned. The main objective is to widen market outlets and make these operations economically profitable.

The new technologies have also led to profound changes in the planning/decision-making level concerning

—the shortness of the average life-cycle of products, which demands extreme flexibility in production and commerce (as to the reduction of the life-cycle of microelectronic components , see Fig. 3);

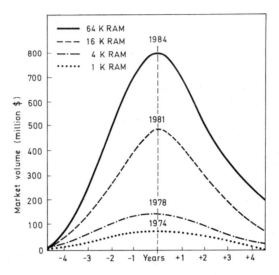

Fig. 3. Changes in the life-cycle of random access memory (RAM) due to the rate of integration. Source: *L'Expansion,* 19 September—2 October 1980.

—the continuous appraisal of the "make or buy" policy to prevent disproportionate internal cost levels or delays in the "internal" availability to key products or key system parts (e.g. peripherals) from compromising the success of an entire range of products or systems;

—the need for links with other manufacturers to exchange know-how, patents, licences, or to sell/acquire components, finished or semi-finished products;

—the location of industrial sites which no longer depends solely on market considerations or on comparative labour cost advantages, but also on proximity to know-how and component supply centres;

—a change in production processes, with a move away from assembly lines for mass production to small and more flexible integrated assembly units for more complex and more frequently varied production;

—a diversification towards complementary mechanical and electronic activities in order to absorb overmanning and improve profitability, by creating a network of small enterprises highly specialised in specific fields;

—a reduction of production integration by decentralising production into small dependent or independent companies in order to increase flexibility in adapting to current technological change;

—a greater commitment to preproduction areas (R & D) and post-production areas (applicative software, marketing, technical assistance) than to production itself (with the previous mechanical technology, investment at the production phase was at least equal to commercial investments, while there was notably less invested in R & D).

This example shows how the impact of microelectronics, which can rapidly lead an enterprise into crisis if passively endured, may be a stimulating opportunity to completely renew company organisation and strategies, and also to recover some of the manpower and value added which would otherwise be totally lost.

Technological conversion (from mechanics to electronics, from an old to a new generation of electronics) is possible only if those involved understand and fully accept it.

Agreements between companies and unions

In the above case, conversion at the shop floor level was made possible by annually revised specific agreements between the company and the unions, and by constant collaboration between management and employees.

The need for such co-operation is also demonstrated by the experience of a major European telecommunications equipment company. In four years, the company passed from the production of electromechanical switches to fully microelectronic switches. Although at first sight the best choice seemed to be to run down its old plants and open new purpose-built factories with an entirely new work force, the company preferred the course of conversion, persuading line managers to act as key agents of the changeover.

A retraining programme was held in which each worker took part for about three months. The flow of work on the factory floor was completely rearranged, taking into account that with microelectronic technology people must have a wider scope of activities than on the traditional production line.

One of the main difficulties was the fact that in the fast-moving world of microelectronics new types of components become available almost every week and project managers tended to revise designs many times, creating difficulties for production start-ups. Finally, it was decided to make changes only at regular and controlled intervals. The conversion programme was very successful, enabling the company to keep ahead of other firms in innovating.

Ability of management to readjust

Experience so far demonstrates that a decisive factor in reacting positively to microelectronics is the ability of management and staff as a whole to adjust. This adjustment can be achieved through broad retraining programmes which must be periodically repeated as technology develops.

It is inevitable that in the future the emphasis will shift still further from production to pre- and post-production phases, i.e. to R & D, engineering, software design, application design, marketing, maintenance and customer assistance services.

This will mean being able to pass from a manufacturing-oriented mentality to a more customer or service-oriented mentality.

The move from a hardware-oriented approach to a software-oriented approach will become even more apparent.

In hardware production, large economies of scale and automated mass production are possible, enabling costs to be constantly reduced.

The software problem

Software production is still mainly the work of craftsmen, essentially based on the individual skills of analysts and programmers. The job of writing software is extremely labour-intensive.

Software developments are not keeping up with hardware for two reasons: the difference in production methods and the short supply of people who can write software.

The difference in production methods corresponds to a difference in costs: the production of a programme line costs on average about $10, which is considerably higher than the cost of an entire microprocessor. Software absorbs over 50 per cent of the R & D expenditure of manufacturers producing major components, and this figure is expected to rise to as high as 80 per cent.

It is foreseeable that in the future a considerable amount of software production will be automated. However, there is no doubt that for a long time to come the ability to produce software through the appropriate use of the available human capabilities will be a decisive factor affecting the chances of electronic companies to adapt successfully to the impact of microelectronics.

Impact on Manufacturing Processes

The factory of the future

The prototype factory of the future is already being designed and there are different prototype models in some industrialised countries using the most advanced developments of microelectronic technology.

These are completely automated factories able to turn out small production batches of different items on the same machines: small cells of numerically controlled machine tools, fed by robots, are controlled by minicomputers, linked in turn to a central computer regulating the flow of work and materials. Quality controls and routine maintenance are handled by robots: any breakdown can be eliminated by the main computer which reorganises the overall plant production schedule. It is possible to link the main computer with another computer containing detailed product designs produced through CAD (computer aided design) technology.

These factories, never completely unmanned, require three levels of highly trained engineers: planners at the overall programme level, computer operators at the computer production level and technicians at the workshop level.

An example is a project for a steel mill designed in Sweden. This factory can be run by a team of some twenty people who monitor the automated system, handle clerical work and some of the maintenance. There is close co-ordination in the team between physical handling operations and clerical work for planning and dealing with suppliers.

Apart from prototypes and designs for factories that cannot be put into operation before the late 1980s, in practice the introduction of completely automated manufacturing processes, incorporating microprocessors, is very limited.

Automation has been introduced mainly in continuous processes (such as chemicals, steel, cement, paper, etc.) where it is based on traditional large computers. Whereas now, new distributed control systems based on microprocessors located at all the main points of the shop floor can greatly increase plant operativeness.

The new "steel-collar" workers

The application of microprocessors to numerically controlled machine tools has reduced costs, dimensions and weight, increased multifunctional capabilities, simplified operating methods and reduced maintenance. Multipurpose turning and machining centres, capable of changing tools automatically, have appeared on the market. With the introduction of microprocessors into the industrial robots, the possibility of completely automating manufacturing processes has increased.

Enterprises have high expectations of future developments in robots and their potential for use in manufacturing activities (namely in joining and assembly processes which are extremely important in terms of quality and product costs) and possibly also in other areas like materials handling, product transportation, packaging, etc.

Cost/precision factors have limited the diffusion of industrial robots. In the future, however, with the cost reductions expected in microelectronics (and also in sensors and actuators which account for a large proportion of the cost of a robot), they will be introduced in a massive way. By reducing robot costs from $50,000 to $10,000, shipments of robots in the United States could reach 200,000 units annually by 1990.

The introduction of robots is intended to minimise assembly costs, increase productivity and improve quality. It has been calculated that in one year, for every $1 spent on robots, $3 are saved on other costs.

Major developments of what have been called "steel-collar workers" are forecast for the 1980s, leading to the widespread elimination of all hazardous and tiring production processes and of most assembly work. But, together with the benefits which will accrue from the improvement in working conditions, there will be a sharp reduction of blue-collar workers.

Large robot lines will be introduced extensively in the assembly lines of the automobile industry: this seems to be the only way for the American and European industries to competitively face the challenge of the Japanese car industry which is already highly automated.

In a Swedish car factory twenty-nine robots at a cost of $7 million are performing the work previously done by 70 men in two shifts. A Battelle investigation (also referred to in the chapter "The Worker and the Workplace") on the effects of robotisation on employment showed that per robot 1.5 employees per shift could be saved, while tool handling robots (for welding work) gave a saving of 0.9 and workpiece handling robots (for forging presses) about 2.1 (mostly un-skilled) people. On the other hand, each robot used creates 0.3 jobs in fitting, pre- and on-the-spot programming and maintenance.

Because the technology is highly flexible and adaptable, microelectronics is decisively affecting the evolution of robots and their diffusion. In the past, older generations of electronics did not offer such a degree of flexibility: consequently automation was limited to more rigid manufacturing processes. But enterprises now realise that microelectronics puts at their disposal an extraordinary range of machines equipped with "brains", "memories", "eyes", "ears" and every kind of sensor and actuator, at very limited costs. As a management survey of 100 companies in Japan shows, the applications of microelectronics in production processes attract greater interest than improvements in administrative work or the ability to develop new products, which are also possible with microelectronics.

A new kind of productivity

The main objective appears to be improved productivity through the substitution of human labour by robot labour. The time of productivity improvements based on the constant growth of production volume appears to be over, and so increases in productivity are closely linked to the capital invested per product unit.

In practice, various companies realise that the spread of microelectronics in manufacturing activities is not simply the replacement of human labour with robots, but a complex phenomenon which demands the total reorganisation of the production processes and creates a new type of value added. This new value added is the set of information which microelectronics generates as a by-product of many automated processes. This enables production to be integrated into the other company processes such as automatic design, automation of order handling, materials handling, packaging and warehousing, through the use of CAD (computer aided design) and CAM (computer aided manufacturing) technologies.

The flexibility and programmability of manufacturing systems incorporating microprocessors make it possible to reorganise manufacturing work in all sizes of production units, ranging from large plants to small and very small plants, which can easily adapt to frequent changes in products and product volumes.

The entire factory layout and work organisation must be redesigned and restructured if maximum benefit is to be obtained from the new systems. These changes are sometimes opposed by line managers and production engineers on the one hand, and by the workers' unions, concerned about employment cuts, on the other. The introduction of automated systems, if imposed, often leads to only partial improvements in productivity, inferior to those possible when the process of innovation is completed and accepted.

A problematic aspect concerns the type of personnel needed when new automated systems are used: there are cases, e.g. robots used for loading where semi-skilled operators are sufficient, with a reduction of average qualifications; in other cases, particularly where more highly integrated systems are used, a particularly high level of technical skill is required for monitoring and maintenance. To sum up, the applications which have been tried out so far are extremely variable, and allow only a few conclusions to be drawn about the great revolution which microelectronics seems to bring to factories in the coming years.

Excessive slowness or excessive speed in introducing the new automated manufacturing processes, without the effective reorganisation of the factory and without the agreement of those who will be working in it, can have disturbing consequences for both employment and quality of labour, as well as for the development prospects of enterprise activities.

Impact on the Service Enterprises

Banking and insurance

Until now the most significant applications of microelectronics in the service sector have been developed in banking and insurance activities. The use of teller terminals, automatic cash dispensers and EFT (electronic fund transfer) systems within and among banks has developed at significantly higher rates than in other user sectors.

The insurance sector has intensified the automatic handling of policies. In the future, banks will be a particularly suitable area to try out office automation applications and the new integrated networks for combined data and text processing. Some important projects being jointly developed by systems manufacturers and major banks or bank syndicates are good examples of this. So far, automation in this area has had no negative effect on employment, because of the very strong growth rates of banking activities.

However, this growth is not likely to continue indefinitely and may even reverse itself, given the bleak prospects for general economic development and the intensification of automation over the next few years.

The organisational structure of banks, and particularly of local branches, will undergo radical changes. Microelectronics will in fact make low-cost and suitably protected cash dispensers and terminals available for direct use by the public, even for complex banking operations. It will be possible to carry out these operations not only in the branch office, but in public places such as airports, department stores and even, in the future, in private homes, thus extending the bank structure beyond its premises. Many of the tasks currently performed by the branch offices will thus disappear and they will, instead, have to develop new customer services, providing financial, legal and fiscal information and advice.

There will be substantial changes in bank management as well, which will be less concerned with routine tasks and much more with providing financial advice and assistance for customers (enterprises and investors).

The evolution towards a cashless (and increasingly chequeless) society, where the flow of money will turn into a flow of electrons around the circuits of computer networks, will radically affect not only the organisation of bank services but commercial services as well.

Commercial services

In commercial services through point of sales terminals (with label scanner) and on-line accounting systems, through total automation of inventory control, of product pricing and restocking, microelectronics tends to favour the concentration of sales structures in department and chain stores, leaving less room for the small independent retailer. Consequently there is likely to be a reduction in the number of small commercial firms and an increase in specialisation among those remaining.

However, microelectronics makes small, low-cost systems available to the small commercial units as well, which may ensure the same business efficiency as the large department store has. The basic ambivalence of microelectronic technology is apparent in this field as well, for while it leads to a concentration of enterprises, it can also help small structures to develop through intensive and specialised application.

In the future, as home terminals become more widely used, enabling the consumer to make his purchases without going out of the house, commercial enterprises will have to carry out extensive reorganisation programmes; as will the newspaper industry and other service areas.

Communication services

One of the areas in which microelectronics has a major impact is telecommunications, where it extends the field from the telephone service to data, text, graphic and image services. The monopoly of the postal, telegraphic and telecommunications authorities in telecommunications in nearly all industrial countries will be challenged by alternative private services, as in fact is already happening in the United States. The public bodies will thus be forced to make radical reorganisational changes in the running of their services, introducing new marketing and promotion techniques, new tariff policies and technical assistance.

As the protective barriers of the public monopolies in telecommunications (and in postal services as well) break down with the development of new microelectronic capabilities, conditions will favour the growth of strongly innovative enterprises in these sectors providing specialised service.

Other new areas which will experience spectacular growth as a direct result of new microelectronic applications include information services, cultural services and entertainment.

The decreasing cost of videodisc equipment, the use of home terminals for access to information banks, the automation of libraries and the spread of computer aided education are stimulating the development of a large number of companies in these enormous markets. The large financial and publishing groups are also turning their attention to these areas.

Electronic information is a sector where the current entrepreneurial approach to the production of cultural information services (books, newspapers, radio, television, etc.) needs to be rethought to incorporate new types of economies of scale.

In fact the possibilities opened up by the new technology may lead to a total standardisation of information and "cultural goods". Alternatively they might lead to an effective personalisation of the products offered to individual users. Both these paths are open and the information suppliers can follow them. Here as well it is a question of making adequate provision, through regulatory measures if necessary, for a valid exploitation of the "customisation" made possible by microelectronics.

MAS - K

Impact on Administrative and Office Work

Structured versus unstructured activities

In the administrative environment certain activities may be defined independently of the person executing them. These activities, which may be called structured activities, include for instance cost accounting, invoicing, payroll, order processing, finance, inventories, etc. Structured work is characterised by fixed procedures, routine tasks, predetermined sequences, codifiable data. There is a clear link between this type of work and the computer's programmed method of working.

Structured activities in clerical work have been automated, mainly in large companies, following the successive developments in data processing: from first generation computers, which were not unlike traditional accounting machines, to second generation ones using a more integrated systems approach, and finally to interactive computers in distributed rather than centralised configurations.

Microelectronics is causing a major increase in the automation of structured jobs and an expansion of the area of structured work because it makes low-price computing power available to the various levels of the enterprise organisation. It facilitates the decentralisation of computing power and improves access to information.

Structured activities represent only one third of the administrative and office work in an enterprise.

The other two thirds may be considered unstructured activities, such as managerial/secretarial work, information flow via telephone conversations or written messages, meetings to take interactive decisions, non-routine exchanges of documents, etc.

In this area traditional data processing has achieved very few results, and increases in productivity and efficiency have been very limited. With the new opportunities created by microelectronics it will be possible to extend information processing systems to a growing number of unstructured activities in the so-called "office automation" area.

Change in traditional jobs

According to a British clerical union survey, a word processor can raise a typist's physical productivity by 30—80 per cent by reducing time spent in retyping, correcting and handling paper.

In the international money transfer department of a U.S. bank, 50 people handling transactions on computer terminals can perform the work done by 430 people ten years before. A U.S. firm is designing a complete teleconferencing network by means of which executives in different locations can confer with one another visually by use of satellite hook-ups; the company claims that more than $50 million annually can be saved in travel costs.

In another company the use of an electronic mail network to send messages to and from different company locations has markedly reduced the time previously spent in typing letters and sending them by mail.

It has been calculated that so far investments in office automation have been extremely low: only $2000 capital investment per employee, compared to $35,000 capital investment per factory worker. For this reason there is great potential in the office automation area for the introduction of microprocessor-based systems, particularly "smart" electronic typewriters with communications capabilities, video terminals and "smart" telephones (i.e. with voice and digital input capabilities).

The use of integrated word processing systems, the spread of public or private electronic mail systems, the appearance of a terminal on every office desk will cause radical changes in the companies' organisational structures. In the past whether centralised or decentralised processing systems were used in accounting and administrative work, it caused major changes neither in the enterprise organisation nor in employment. Computers somehow became joined to traditional jobs, without drastically modifying or replacing them. This is not expected to be the case with the impact of microelectronics on office automation, as indicated in the chapter "The Worker and the Workplace".

The first documented experiences show that the new systems for automating office work are generally regarded by companies as a means of saving labour rather than of increasing efficiency, in other words, the approach is towards saving employee costs rather than obtaining particular end-results. There is clearly great need to rationalise jobs and reduce the time now wasted. The amount of time wasted in clerical work is considerable and is growing: 30—50 per cent of secretarial time is spent in waiting, trying to make telephone calls, retyping, etc. The same holds for telephone calls, where less than 25 per cent of the calls attempted reach the desired party, while the rest encounter engaged signals, no answers or equipment failure; so the time wasted is enormous and telephone channels are idle 80 per cent of total time.

Thus word processors are seen by enterprises mainly as technological devices capable of doubling secretarial typing time, and electronic mail is seen as a means of eliminating secretarial time wasted on calls or even the secretarial job itself. But there are examples which show how new systems can increase the quantity and quality of activities in the office, without reducing the number of jobs involved.

What is needed is that the enterprise should use technological change as a driving force to restructure and reorganise administrative work, especially at the management level where microelectronic systems can help to transform decision-making into a swifter and clearer process.

The "paper explosion"

The quantity of paper documents circulated in the administrative and management area is proliferating in an exponential way. A great deal of bureaucratic work is performed to cope with the paper explosion and this work, in turn, generates still more paper. Automating the handling and storing of documents is fundamental to the rationalisation of this process which is growing out of control. But the enormous quantity of data and information that can be filed in the new quasi-zero priced microelectronic memories can make decision making in companies even more difficult unless new Management Information Systems or Data Base Management Systems able to make rational use of these enormous resources are developed. Several examples show that where electronic information filing is simply a substitute for manual filing, with no suitable provision for automatic connection with planning, decision making and control functions, it can lead to a deterioration of efficiency.

Work improvement or control?

From the employee's point of view, a terminal on every desk may also mean controlled and constantly monitored work, diminishing work content, conditions of stress, and fewer human contacts. Thus this solution may be rejected by both employees and management, as has in fact already happened in some "total office automation" experiments. Everything depends on the way in which this process is handled. If the design of the new systems for administration and for the office takes into account the need for qualitative improvement in the work content in these areas, then real progress will be made in raising the productivity and standard of these activities, with benefits for both the enterprise and the employees. The reorganisation of enterprises through the automation of administrative and office work must take place with the participation of those involved in these activities, both management and clerical staff.

Automation of the office is quite different from the automation of the factory. In the managerial and administrative area, the expansion of microelectronic automation may lead to the elimination of many functions and increase the need to learn new skills and to readapt. Hence, mobility both inside and outside the enterprise will grow enormously and it may be necessary to develop forms of part-time employment.

Despite the speed with which microelectronic technologies spread, the process of change appears to be slowed down by management's reluctance to implement drastic changes which could have a negative effect on their own future.

Impact on the Overall Enterprise Organisation

High investment in R & D

It has already been emphasised several times on the preceding pages that the diffusion of microelectronics in enterprises creates need for major restructuring, not just at the manufacturing stage, but at all levels. In the electronics industry, in particular, there is a growing need for high investment in R & D, which is of fundamental strategic importance (as is the manufacturing stage in the mechanical industry). The speed of microelectronic change is such that innovation in enterprises wishing to survive proceeds at a frantic pace.

The life-cycle of components and of the products incorporating them tends to be very brief, and only those companies which manage to keep ahead of their competitors can be sure, for a couple of years, of margins high enough to cover the heavy reinvestment needed in R & D. After a year or two imitators enter the market and prices and margins drop rapidly. Therefore only if a company is able to lead innovation in its specific product, and thus be several steps ahead of its rivals, can it achieve the level of earnings needed for reinvestment in R & D. And vice versa, to be a leader of innovation, sufficient resources need to have been invested in R & D. This is a circular process which creates extremely high competitiveness and accelerates innovation at a rate never experienced in other sectors.

As component integration in information processing systems increases, the hardware-software architecture must be completely redesigned; but companies cannot afford to wait for the new products based on them. They must exactly forecast when the commercial production of the next generation will start, and then design their products to coincide with the availability of the new components. The risk of miscalculating is high and increases as the degree of integration grows.

"Make or buy" policies

Faced with the speed with which microelectronic developments occur, companies incorporating components in their products have developed complex "make or buy" policies, not only for components but for intermediate and end-products as well, to ensure maximum flexibility and adaptability to new products: it is normal practice for a company to add other companies' products (e.g. peripherals) to its own for temporary periods in order to retain market shares, even if its own production structures are sufficiently flexible and rapid in introducing new models.

Purchasing policy thus assumes great importance, since it makes it possible to reach an adjustment speed and cost levels which internal production often does not allow. There is also a tendency to make use of several supply sources for microcircuits and software packages. Even the large systems companies which manufacture in-house microcircuits tend to use external sources to reduce the risk of a delay in availability. This causes still greater variations in trends in value added and in production manpower requirements.

Company planning must constantly bear in mind the need for technical and manufacturing agreements, exchanges of know-how, patents and licences with other manufacturers, or mergers and acquisitions.

The extremely complex relationships which are thus created among the various companies highlight the highly international nature of microelectronic technology. Enterprises which limit themselves to a single market and do not expand internationally are finding it hard or impossible to survive.

Effects on marketing

Commercial and marketing structures are also greatly affected by the spread of microelectronics. Planning in marketing becomes more closely associated with product planning and the design stage. There is increasing interaction between technological development and the application requirements of the market.

As electronics enterprises become less manufacturing-oriented and more service-oriented, it becomes necessary to make significant changes in marketing and after-sales service. The fact that the new information processing systems, telephone switches, automated manufacturing systems can incorporate microprocessor with self-diagnostic and self-repair functions, eliminates the need for traditional maintenance workers.

There is, instead, a need for personnel able to assist users in installing and subsequently operating equipment and systems. The sales and customer assistance area is thus developing in a direction away from a product-oriented approach and towards a service approach.

Need for training

This adjustment can create big problems and cause tension within enterprises. The new skills required differ greatly from those which are available. Retraining experience has shown that there are limits to adjusting, limits related to age and educational level.

The same problem occurs at the managerial level as regards the new managerial techniques based on office automation approaches, and the changes in planning methods which microelectronics makes necessary.

Positive adjustment without economic and social upheaval seems to be possible only if people can be trained and retrained according to the criteria required.

Some companies understand that the impact of microelectronics will be positive only if people have an adequate level of knowledge for making the best use of the advantages of microelectronics. So they are allocating considerable sums to training which, together with investments in R & D, constitute the real "capital formation" of the microelectronic era.

5

The Worker and the Workplace

JOHN EVANS

Introduction

The most immediate question facing working people concerning microelectronics is whether the introduction of new technology will result in a loss of jobs. The following chapter on "Microelectronics and Macroeconomics" indicates the scale of job displacement that is likely to arise over the next ten years due to the introduction of new technology based upon microelectronics. However, equally important is the effect that microelectronics will have upon the nature of people's jobs and on the environment in which they perform them.

The application of microelectronics to products and processes has already produced some substantial changes for workers in industries such as precision engineering, motor vehicle manufacture, printing and parts of the service sector such as banking and insurance. The changes can often be seen to have conflicting effects upon the quality of work. Some jobs are eliminated, others are deskilled and yet others are upgraded. In some cases, the use of microelectronics can lead to an improved physical working environment, whilst in other cases it can lead to problems of increased isolation of the worker, increased supervision of work and the changed pace of work.

These conflicting effects reflect the fact that a particular technology such as microelectronics does not determine a particular form of work organisation or a set of working conditions. It is rather the economic, technical or social objectives which the technology is used to achieve that determine the effects of the technology upon working conditions. If a rational assessment is made of the longer term economic and social choices presented by new technology, then certain positive opportunities for improving the quality of working life are given by

157

microelectronics. If, however, the driving force behind technological change is only one of increasing productivity and short-term profit, then this may lead to a general deterioration in the quality of work. This has been the motivating force behind much past and present technological change: where social factors have been taken into account, this is only normally due to economic reasons such as the difficulty of recruiting staff to work in unpleasant jobs or work places.

There is a danger that against the present background of economic crisis and high unemployment, the rationalisation possibilities brought about by technological change will be pursued to the exclusion of all else. Consequently, the current wave of technological change based upon microelectronics is in danger of precluding a series of technological options.

The Quality of Work

Defining or measuring the qualitative aspects of work is a highly subjective process; nevertheless, in practice workpeople do make judgements about satisfying and dissatisfying aspects of work and considerable experience has been built up in analysing what determines job quality. This suggests that the main determinants of job quality are: job content; its meaningfulness; the learning involved; working environment; job security; social contact; rewards; and the impact upon leisure time. Important factors affecting job content include the skill requirements, the responsibility, the freedom from supervision, the control over work pace. In determining a job's meaningfulness, it is important for a worker to see the role the job plays in the overall production process and see its relevance to the final product.

The application of microelectronics affects all of these aspects of work, whether the application is in the form of a change in the process of production or provision of a service or whether it is through a change in the end product. This chapter assesses the impact of microelectronics on these qualitative job features before going on to consider the technological alternatives for the future.

Microelectronics and Skills

The chapters "The Technology" and "The Technology Applied" have indicated the potential that microelectronics gives for product and process applications. A number of conflicting effects on skill levels can be seen in these applications. On the one hand, a certain number of jobs are completely eliminated and the skill requirements for others are reduced. On the other hand, however, the skill requirements for certain categories of jobs are increased. These three effects, job loss, deskilling and upgrading, are part of the process of change.

Microelectronics can be used to eliminate many unskilled and repetitive jobs. It can also be used to standardise and deskill qualified jobs. The standardisation leads to a concentration of required skills in a limited number of new jobs which are created essentially in the planning and production service areas. These new skilled jobs are frequently computer-related.

The Elimination of Unskilled Jobs

Surveys that have been carried out to assess the factors that are conducive to the application of microelectronics have shown that it is particularly suitable in automating industrial processes where existing jobs involve tasks which are repetitive or sequentially repetitive. Much of the development of industrialised societies has involved the creation of such lowly skilled jobs, through the division of labour, specialisation of tasks and mechanisation of production processes.

This has not simply been the result of mechanisation through technological change, but rather the application of scientific or Taylorian management theories. Systems of production based upon these theories have been characterised by the separation of the planning and execution of work, the breaking down of work into its smallest and simplest parts and rigid management control over the production system. The "line" production system is the norm in large parts of manufacturing involving mass production, although the introduction of alternatives to line production systems is becoming increasingly common.

A West German study has estimated that in 1977, 33 per cent of men and 46 per cent of women employed carried out repetitive jobs with low responsibility, whilst 11 per cent of men and 40 per cent of women carried out jobs involving simple auxiliary tasks. The potential for the full automation of these tasks is greatly enhanced through microelectronics.

An example—robotics

An example of such an application is the development of robotics, in which straight substitution of human by mechanical activity takes place. The chapters "The Technology", "The Technology Applied" and "The Impact on the Enterprise" have referred to the technological advances now taking place with the development of robots and to their range of applications. Whilst the main use of robots up to now has been in the mass production industries such as motor vehicles (where the majority of American, European and Japanese producers have introduced robots), metalworking, chemicals, civil engineering, electrical and electronic industries, their use will expand over the next ten years into wider fields of application.

The main tasks carried out by robots are those of assembly, joining and handling. A major American manufacturer of robots has estimated that 50 per cent of the machines it has installed have been used for spot welding, about 11 per cent for dye-casting and another 5 per cent for machine loading. With advances in the application of microelectronics, particularly in science or technology, universal machines are being developed which are able to carry out more complex assembly tasks.

The study on the introduction of robots in the Federal Republic of Germany, referred to in the chapter "The Impact on the Enterprise", examined the introduction of ten robots in five companies in a variety of sectors. The tasks affected were arc welding, paint spraying and handling of pieces. The effects on manning requirements are shown in the table below. A total of 46 people in non-qualified jobs were affected; of these 7 were made redundant, 28 moved out of the sector, 10 changed jobs within the sector, and one moved into the sector. The introduction of robots therefore meant a substantial change in the jobs that workpeople did.

Staff fluctuations due to the introduction of industrial robots:
a study of the introduction of ten robots in five West German companies

Category of personnel	Changes in work tasks					
	Staff moved out of the sector	Staff moved into the sector	Change of staff within the sector	Staff leaving	Recruitment	Total
Qualified staff	4	6	2	—	—	12
Non-qualified staff	28	1	10	7	—	46
Total	32	7	12	7	—	58

Source: Battelle, Geneva.

A different West German study of 40 applications of robots in a major motor vehicle manufacturer found that one robot on average replaced 4 existing workers, but led to the creation of one new working place. One third of the new jobs created were of a lower quality than the jobs they replaced due to a loss of job content.

The positive aspects of the application of robotics have included the elimination of a range of unskilled and repetitive jobs. Studies on the impact have also shown a reduction of strenuous handling work and the possibility of a reduction of jobs in dangerous or unpleasant environments, such as paint spraying. Trade unions in a number of countries have used the changes in line production due to the introduction of robots and automated materials handling as a means of negotiating changes in work organisation in the interests of improving the jobs that remain even in a highly automated workplace.

The Deskilling of Jobs

The removal of unskilled jobs due to automation has, however, been matched by a downgrading of skilled jobs as a result of the application of microelectronics.

An example—computer numerically controlled (CNC) machine tools

The deskilling process also leads to a change in what is understood as skill. In manual or mechanical technologies, skill is frequently experience-based, requiring long period of apprenticeship or training to acquire. This is now shifting with the advent of new technology to skills based upon analytic or logical ability rather than experience.

One example of deskilling has taken place in the industries using machine tools for small batch production. Numerically controlled machine tools were first developed in the 1950s, whereby a machine's operations were controlled by the use of programmes of binary digits fed into the machine via paper or magnetic tape. With the reduction in the cost of computing due to microelectronics, computer numerical control has been introduced, whereby a minicomputer controls a machine or a group of machines and is programmed according to the specific job that is required.

In most industrialised countries the operators of conventional machine tools in small batch production have traditionally been highly rated within the hierarchy of manual jobs. The task is highly skilled and requires a long period of training. The tasks of the machinist include assisting with the planning of the production of a part, setting up the machine and controlling the speed and operation of the machine according to varying local circumstances such as the quality of the metal. Whilst the job involves communicating with both management and the draughtsman and designer of a part, the machinist retains a high degree of freedom and control over his job.

The introduction of CNC machine tools can be seen as a process innovation which is generally substitutive in its effects on labour skills. The skills of the machinist are broken down into their logical components by the analyst and programmer and the computer is programmed to control the operation of the machine tool in a way similar to the machinist. The control of the machine tool is transferred from the operator to the systems specialist, who analyses his skills. The job of the machinist is changed into one of monitoring the computer controlled equipment—a role that will largely be eliminated once CNC systems become more reliable.

An important development will be the link of computer controlled machine tools to computer aided design (CAD). This will also eliminate the liaison between the machinist and the designer in the planning and design process. The tool-setting can be prepared at the design level, with the computer setting out the production and engineering specifications for the machine tools. The tasks of the machinist after a time, therefore, tend to become redundant in an automated machine-shop.

Some commentators have argued that the deskilling process, by standardising previously skilled jobs, is thereby preparing the way for the next phase of automation which will eliminate such jobs. Some countries are currently conducting research programmes with the objective of making operational an unmanned factory for the production of engineering parts.

The impact that the introduction of CNC machines has upon skills can, however, vary, depending on the circumstances in which it is introduced and the ends to which it is put. Where CNC is introduced to provide increased flexibility and design sophistication in the product, a highly centralised form of work organisation may not be satisfactory. Different organisational options do exist with the introduction of CNC machines. Due to the reduction in cost of computing brought about by microelectronics, data processing facilities are now distributed to the level of the machine tool. This means that operating programmes can be edited and altered at the level of the individual machine by a system of manual data input. Under centralised systems of CNC machine tools outlined at the beginning of this section, any editing of programmes would be carried by a computer programmer from a data processing department. In practice, however, it is possible for the machine tool operators to carry out an editing function if trained in basic computer programming skills. Indeed in cases of small batch production or relatively straightforward production requirements, it may be possible for CNC machines to be programmed at the shop-floor level. Besides leading to a more flexible production system and a better end product, such alternative options also lead to increased skill requirements in the machinist's job by harnessing existing mechanical skills with new computer skills.

In both Norway and Sweden trade unions have negotiated in particular plants to ensure that machinists were given supplementary responsibilities for the programming of CNC machine tools. This required the retraining of the CNC operators to give them computing skills and the redesigning of the organisation to give the shop floor access to the computerised production and inventory system. In one Norwegian plant, five machinists were given a year's training in computer programming to re-equip them for a CNC tool shop.

In the United Kingdom, trade unions have reported that in some cases machinists have picked up the skills of computer tape editing from working on CNC machines and talking to the engineers involved in their installation. This was achieved without any formal computer based training, although it led to subsequent demand from the work force for an expansion of training and a distribution of the programme editing function to existing machinists.

An example—printing

A further example of an essentially substitutive form of process innovation leading to deskilling can be seen in the printing industry with the switch-over from a mechanical system of typesetting based upon hot lead, to computerised typesetting based on photocomposition. This has been made economically feasible by the reduction of cost of computing as a result of microelectronics. Microprocessors are also coming to be used in various ancillary control functions in the printing process such as press and ink controls.

The newspaper industry in most European countries, faced as it is by difficult problems of viability, has come to be the focus of much attention for the application of new technology. Computerised typesetting allows the preparation, editing and final layout of text to be carried out on a visual display unit (VDU).

In one example of the change in the level and pattern of employment as a result of the changeover to computerised typesetting, the total number of jobs on the late shift in a newspaper printing plant fell from 52 to 29. The jobs of 12 typesetters were replaced by 5 VDU operators involved in work preparation, the jobs of 10 correctors were replaced by a further 5 VDU operators and 15 jobs involving layout were replaced by 7 jobs. The jobs involved with typesetting, correcting and layout were all essentially highly skilled jobs involving high remuneration, a high degree of personal or group control and requiring an extensive period of apprenticeship or training. They have been replaced by essentially lowly skilled VDU operating jobs requiring only basic typing skills. In some areas specialised operators have been removed completely as a result of "direct entry" of copy into the computerised system by journalists. New skilled jobs are created, however, in the data processing field.

The increased dangers of deskilling in the newspaper industry as a result of technological change, together with the traditionally strong position of those groups affected, has manifested itself in a series of industrial disputes in the newspaper industries of several countries. There was a major dispute in the printing industry in the Federal Republic of Germany in the winter of 1977/78 over the treatment of displaced typesetters and printers. The question of downgrading was a central issue in this dispute. The final settlement gave an eight-year job guarantee for skilled workers with six years guarantee of wages comparable to the former status. In Sweden, an agreement has been reached providing for a full job guarantee for displaced workers in the national newspaper industry.

In the United Kingdom the problems of loss of jobs and deskilling came to a head in the dispute between the British print unions and *Times* newspapers, which resulted in a lock-out of the workers at the *Times* group from November 1978 to October 1979. In the United States the *Washington Post* introduced computerised typesetting without union agreement. During the ensuing industrial dispute, 25 managers operated the computerised system, taking over the jobs of 125 typesetters.

An example—clerical work

The chapters "The Technology Applied" and "The Impact on the Enterprise" have pointed out that microelectronics will result in the application of electronic data processing more broadly across the service sector, affecting a range of services not previously susceptible to automation. Particular attention has been drawn to the impact that microelectronics is having upon clerical work. It is in this area where it is likely to have a predominantly substitutive effect upon labour input by reducing the cost of computing and allowing the automatic processing of textual information traditionally performed manually. Word processing systems are currently being sold in substantial numbers in most industrialised countries and their impact upon clerical jobs is likely to be felt increasingly over the course of the 1980s.

One of the main clerical occupations is that of the secretary. Secretarial work consists of two main activities: correspondence work including a large amount of typing; and administrative work. The exact configuration of the job is likely to vary according to the organisation of the work, depending upon the proportion of time spent in carrying out correspondence or administrative tasks. In general, administrative tasks are ranked as being more highly skilled than straight copy typing, although copy typing does also require skill in controlling the layout and quality of work.

The introduction of word processing systems affects the skill levels of secretarial work in two ways. Firstly, it can result in a change of work organisation with a splitting of typing work and administrative work. Secondly, it changes the skills required by the copy typist.

The introduction of word processors is frequently done through the establishment of a separate word processing department which would receive the bulk of typing work within an organisation. This allows machinery to be kept working more intensively than is the case with traditional secretarial work structures and so achieves the maximum use of the capital equipment. The changed work organisation results in the "deskilling" of the word processor operator compared to a traditional secretary, since the job becomes one of essentially typing and no longer contains more varied administrative roles. Contact between typists and authors is also reduced as normally material is fed through the supervisor of the word processing department, who would also deal with any queries.

This organisational structure as described is only one possibility. An American survey of the introduction of word processing systems has described four different structures. One is a centrally administered system within an organisation where all correspondence secretaries are located at the word processing centre and all typing within the organisation is sent there. An alternative is to have satellite centres with typists trained in specialist typing skills and housing both administrative and correspondence secretaries. A third structure would have to be a back-up centre as an overload system for traditional secretaries. A fourth structure is to have a decentralised system where word processing facilities are located in the normal departments and less division takes place between the work of administrative and correspondence secretaries. The less centralised the system, the less polarisation of skills takes place.

Word processing also changes the skills of the copy typist. Less skill is required in the layout, accuracy, execution and correction of work, since this is all performed by the machine. However, skills additional to typing are required in operating the machine and a training course of one or two weeks is normally required to train a copy typist to become a word processor operator. Where work involves the typing of standardised texts, much of it can be automated and the copy typist's job becomes even more deskilled. For example, typing standardised letters may simply require coding work.

Case studies have shown that good and experienced copy typists have felt a deterioration in the skill level of their jobs with the introduction of word processors, whereas younger typists have felt an increase in skills due to the chance to operate a more complex piece of machinery. One British clerical union has found, however, that once the novelty of using word processors wears off, operators become disillusioned and feel more strongly the disadvantage of increased work specialisation.

In the introduction of office technology, choices concerning work organisation do exist. The choice of whether or not to set up a highly centralised word processing department responsible solely for correspondence typing, with independent departments containing their own administrative secretaries, is a social decision, not a technical one. An equally feasible organisation is for word processors to be used on a shared basis so that the correspondence content of several secretaries' jobs could be reduced, allowing more time for administrative tasks. This has been achieved in some cases by job rotation.

The economic objective of maintaining maximum utilisation of the word processing machinery may become less important, due to the reduction in cost of word processors in the future.

An example—substitution of mechanics by microelectronics

Examples of deskilling can also be found in the substitution of mechanics by microelectronics. One of the most significant employment effects of the new technology to date has been in those industries manufacturing products in which mechanical or electromechanical elements have been replaced by microelectronic elements. The reduction in the number of components with microelectronics has led to simplification of the assembly process and a reduction in the labour input required to manufacture certain products as pointed out in the chapter "The Impact on the Enterprise". Products radically transformed in this way include telex writers, sewing machines, watches, calculators, cash registers, etc.

In watchmaking, the mechanical and manual tasks required in manufacturing a traditional watch involve more than a thousand separate operations; these have been replaced by the assembly of the five basic components of an electronic watch: battery, quartz crystal, integrated circuit, display and case. As a result skilled jobs have been lost in watch repairing. Another example is in telex machine manufacturing, where in one electronic model microelectronics have replaced 936 parts in its electromechanical predecessor. The change in the product resulted in a fall in the proportion of jobs in the manufacturing process which required training, from 82 per cent to 35 per cent. At the same time, the proportion of highly qualified technical staff in total employment rose from 2 per cent to 30 per cent. In the Federal Republic of Germany, there have been a number of examples of the industrial relations consequences of deskilling in the mechanical engineering sector. In some cases downgrading took place amongst assembly line workers of up to four grades in a total system of twelve grades. In other cases works councils were able to limit downgrading to two grades and in one case strike action took place against downgrading, and a lock-out affecting several hundred thousand workers resulted.

Deskilling has also taken place in the maintenance field as a result of the replacement of electromechanical systems by microelectronic systems. The maintenance requirement of electronic equipment is lower than that for electromechanical equipment due to the fewer components and moving parts involved, and the possibility for self-diagnosing fault-finding features. This can lead to the deskilling of the formerly highly skilled job of the maintenance electrician. Maintenance at one level becomes a relatively straightforward job of replacing faulty modules. At the same time, however, certain higher skilled electronics jobs are created in the maintenance area as is referred to below.

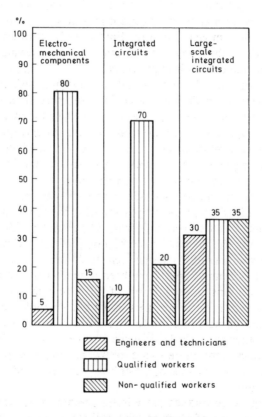

Fig. 1. Electronic components: change in the composition of the labour force due to technical change. Source: Haroldo Corrêa de Mattos, Technology and developing countries in the International Telecommunication Union (ITU), 3rd World Telecommunication Forum, Geneva, 1979, Vol. 1, p. 11.

An example—production of semiconductors

An international survey of electronics firms carried out for the OECD in 1977 found that in semiconductor manufacturing firms skill requirements were rapidly being reduced and the production of large-scale integration (LSI) chips was coming to increasingly resemble a chemical process operation. The report concluded that in general a polarisation of skill requirements was taking place between highly skilled design, test, and inspection personnel on the one hand, and a number of unskilled machine operators on the other, with the elimination of semi-skilled production workers. Many of the firms interviewed in the survey said that the polarisation of skills was marked by the export of unskilled jobs to overseas assembly operations, so that the effect of developments on domestic employment appeared to be the creation of a few highly skilled jobs. Figure 1 indicates the change in the composition of the labour force in the electronic components sector due to the move to LSI.

Upgrading of Skills

It was stated earlier in this chapter that the application of microelectronics can lead to a concentration of required skills in a limited number of new jobs, which are primarily computer-related. The skills, for which there is an increasing demand as a result of microelectronics, can be broken down into three broad types. Firstly, there is an increasing demand for engineering skills, primarily in the electronics fields, but also involving the understanding of control, production and mechanical systems. Secondly, there is an increasing demand for logic, systems and software skills. Thirdly, there is a demand for general data processing awareness, although not specialised.

In the early periods of computerisation based upon the introduction of large mainframe computers, substantial numbers of semi-skilled and essentially repetitive jobs were created in the areas of data preparation and data entry. The development of input and output of information based on the use of keyboards and visual display units (VDUs) has now removed much of the employment of this area with direct entry and extraction of information being carried out by computer users; however, data feeding jobs involving no other activities are still generally low-quality jobs.

The creation of newly-skilled jobs may not be taking place in the same companies or sectors as the loss of jobs or deskilling of jobs. Skilled computer staff are employed in three main areas: firstly, in organisations using computers or microelectronics in their production processes; secondly, in the computer or microelectronics manufacturing sector; thirdly, there has been a rapid growth of computer service companies including consultancy, time-sharing, maintenance, etc.

The Japanese study, referred to in the chapter "The Impact on the Enterprise", considers the employment situation in companies, both manufacturers and users of microelectronics-based equipment. This reported that new, highly skilled jobs had been created in the development and design field in companies manufacturing automated equipment and commercial machinery, whereas both skilled and unskilled jobs had been lost in the companies using this equipment. The same study forecast an overall growth in demand for software engineers of 12 per cent per annum between 1972 and 1985.

A West German study referred to earlier in this chapter showed that the introduction of industrial robots led to a net inflow of qualified staff into the affected sector, whilst there was a much larger outflow of unqualified staff. Similarly, the case studies concerning the printing industry showed the increased employment of qualified data processing staff as a result of the introduction of computerised typesetting, at the same time as the displacement of skilled typesetters and compositors.

The parallel effects of the destruction of demand for existing skills and increase in the demand for new skills can also be seen in the maintenance employment area. As microelectronics makes electromechanical equipment obsolete, so the demand for maintenance electricians' skills falls, as was pointed out earlier in this chapter. Although the maintenance requirement of electronic-based equipment is lower than for electromechanical equipment, there will still, however, be an increase in the demand for maintenance engineers with electronic skills and in particular with a knowledge of pneumatics and hydraulics, given the nature of much automated plant.

Whether or not microelectronics therefore provides an opportunity for upgrading or results in downgrading depends in part upon the possibilities for retraining displaced staff with electronics and computer-related skills. It is clear, however, that the problems of implementing retraining schemes are greater with older workers. The option for early retirement coupled with a maintenance of income levels, may in some circumstances be the best solution for easing the social costs of new technology on older workers. Moreover, structural barriers may well exist between different segments in the labour market and the fact that it is skilled manual jobs which are being eliminated and technical white-collar jobs which are being created may create further immobilities. Differences between technologies are also important as mechanical engineers do not always make the best electronics engineers.

Older workers are particularly affected by the introduction of new technology based on microelectronics. They may find problems if forced to shift from mechanical-based skills to electronics-based skills. Options do exist concerning the qualificational impact of changes, but even so a danger of high and enduring unemployment amongst older workers is particularly real. There should therefore be the opportunity for flexible and voluntary early retirement for older workers on satisfactory income.

The rapid increase in demand for computer-related skills has resulted in a general shortage of such staff despite rapidly-growing levels of unemployment. In the United Kingdom, a government-sponsored study suggested that in 1980 compared to a national stock of 275,000 people with computer-related skills, there was an immediate shortfall of 25,000. There was a shortage of 16,000 computer programmers and systems analysts (equivalent to 15 per cent), and a 6 per cent shortage of electronics engineers in the computer supply industry. Estimates suggest that in France and the Federal Republic of Germany a similar situation exists in the labour market for computer staff.

Microelectronics and Working Conditions

It was pointed out at the beginning of this chapter that a number of factors in addition to skills and qualifications determine the quality of a job. These include the control over the pace of work and work intensity, the possibilities for human contact, the freedom from supervision, the physical work environment and the effects upon health and safety. These factors cannot be totally isolated from skills and qualifications, since highly skilled jobs have traditionally been able to attract rather better working conditions than unskilled jobs. Nevertheless, the application of microelectronics can be seen to have an impact on working conditions independent from its effects on skills. As in the case of skills there may be a number of conflicting effects. The reduction of arduous and unpleasant work in some hostile work environments has to be assessed against the increased stress and pressure of work observed in activities in which man-machine interaction did not previously take place.

Work intensity

Microelectronics leads to an increased capital intensity in formerly labour-intensive industries and services. This may lead to economic pressures to use equipment more intensively with a corresponding risk to working conditions.

The introduction of word processing, as has been pointed out, can lead to divorce between typing and administrative functions of secretarial work. Besides leading to a polarisation of skills, this can also lead to an increase in the work intensity of typing jobs. Firstly, the increased specialisation, of itself, reduces the variety of the work and increases the possibility for management control over the pace of work. Secondly, the use of word processors intensifies the actual process of typing work compared to using conventional typewriters. The use of a word processor can raise a typist's physical productivity by reducing the time spent in retyping, corrections and handling paper. A word processor operator is able to type twice as much new material as a copy typist using a conventional machine, given the same work organisation. This means that the possibility of human contact is reduced and the dangers of social isolation at work and work intensity are increased. It should again be emphasised, however, that if certain preconditions are met in less centralised systems, the introduction of word processing can be used to reduce the amount of time spent in typing and increase the time available for other administrative tasks. Such a change could result in both an upgrading of secretarial jobs and reduction in work intensity. The possibility of organisational options is a point referred to elsewhere in this chapter.

The danger of increased work intensity exists among other jobs in the tertiary sector which formerly were not susceptible to automation or mechanisation. The impact that computer aided design (CAD) may have upon the skilled manual worker as the result of the introduction of factories highly automated from the design stage to the production stage has already been mentioned. The introduction of CAD also affects the job done by the draughtsman or designer. On the one hand the computer can be used to carry out the routine or non-creative parts of the designer's job such as drawing, listing parts, specification of manufacturing techniques and production engineering. This both speeds up the design process and eliminates the repetitive part of a draughtsman's job. At the same time, however, the computer can be used in direct interaction with the designer so as to allow the draughtsman to alter designs continuously by giving new requirements to the computer.

This man-machine interaction can lead to an increase in work speed, a reduction of flexibility and an increase in stress. Commentators have argued that the expansion of computer aided design will lead to a growing alienation and loss of job satisfaction in engineering design, due to the subordination of the operator (designer) to the machine (computer).

A deterioration in working conditions due to problems of man-machine interaction need not all result from an increase in the pace of work, but may even arise from a decrease in the pace of work. For example, a stock order clerk using an on-line information system may frequently have to interrogate a computer system to obtain a piece of information to enable him to make a decision to give an answer to an enquiry. If due to an overload of the computer system a reply is not given instantaneously then the job can become highly frustrating. Therefore, the problem is essentially one of ensuring that the speed of operation of a computer system is compatible with the desired speed of operation of the human interacting with that system. This requires a decision being made at a design stage, but this is an economic or organisational choice, not a technological one.

The desire to make more intensive use of equipment can lead to pressures, not just to increase the pace of work, but also to introduce or extend more intensive forms of work as shift working. In one plant of a European electronics-based company, manufacturing automatic tellers, fairly extensive group working systems have been introduced, yet the only part of the plant where shift working is practised is where robots are used to mount integrated circuit chips on a circuit board. The cost of the capital equipment has led management to keep it working continuously. The introduction of large mainframe computer systems in organisations had tended to involve the introduction of shift systems and a certain amount of night work.

Shift working has clear social and domestic disadvantages and night working in particular may have deleterious effects upon health and safety. The two main effects seen have been sleep deprivation and digestive difficulties. Moreover, surveys of attitudes to night work carried out in a number of countries have shown that the majority of workers dislike night work, and seek to avoid it when possible.

Isolation

Both the increased incidence of man-machine interaction and the extension of shift working due to microelectronics lead to the danger of increased isolation of the individual worker at the workplace. The cases set out above have shown that the introduction of word processors in a typing pool can increase isolation due to increased work intensity and increased noise. Similarly, the operator of computer numerically controlled (CNC) machine tools, as pointed out, may not only experience a reduction of skills but also a reduction of opportunity for contact with other operators, or designers and engineers. The decreased manning levels in semi-automated or automated plants increases the risk of decreased social contact at work. This may be particularly true during the night shift in an automated plant where the manning level may be reduced substantially with maintenance work being done during the day.

As the workplace is a major source of social contact in people's lives, an increase in isolation at work may create increased stress and health problems for workers. It also substantially reduces the potential for collective action through trade unions, since lateral contact is reduced. Studies on the introduction of computer systems have shown that they reduce the possibilities for lateral contact and only allow essentially hierarchical contact with superiors or subordinates.

A further reason for increased isolation may be the increase in homeworking that may arise as a result of microelectronics. Homeworking has been traditionally important in certain industries such as clothing and textiles. For employers, the main attraction of homeworking is financial: there is no need to provide factory facilities; in most cases lower wages can be paid because of reduced ability of homeworkers to take collective action through trade unions; and workers can be laid off much more easily than factory-based workers. Microelectronics make technically feasible the expansion of homeworking in other areas such as clerical work. Computer terminals could be located in the home but linked to a central computer via the telecommunications system. Work can therefore be carried out at home but be received, monitored and controlled in a central office.

If microelectronics is used to expand homeworking and there is a reduction in the provision of facilities to enable women to go out to work, then this may well increase the social isolation of women and reduce the social progress towards equality of the sexes.

The jobs particularly susceptible to automation through the use of microelectronics are those which are repetitive or sequentially repetitive. Such jobs are currently concentrated amongst women workers and there is therefore a risk of an increasing incidence of unemployment amongst women. Social and organisational constraints represent a threat to the movement of women into new, more highly skilled jobs. Moreover, the increased potential for homeworking through microelectronics-based technologies may lead to a transfer of women's employment from workplaces to the home with the resultant dangers of increased social isolation. The problems of women with regard to the effects of technological change, therefore, require particular attention.

Control and supervision

It was pointed out above that computerisation can lead to changed work pace for a number of occupations not previously affected by problems of the man-machine interaction. The very process of introducing machines in formerly labour-intensive areas can, however, reduce the scope for individual freedom in carrying out work. This was seen in the examples of the changed nature of the machinist's and the typist's job where the scope for discretion and the decision-making contact of a job may be reduced as a result of computerisation. The chapter "Information Technology and Society" points to the dangers of increased control over the individual working in an organisation due to new technology.

A further problem is that microelectronics can lead to increased information being held concerning individual people in their role as workers. This may be in a number of ways. Firstly, intelligent terminals allow information to be collected on the employees' performance, such as time spent at the machine, work speed, error rates, etc. For those working continuously at a machine such as cash register operators in supermarkets, word processor operators or computer numerically controlled (CNC) machine tool operators, this can lead to an increase in stress and management control over the work pace. Indeed, a dispute took place in the retail sector in Denmark in 1978 when the commerce trade union discovered that electronic cash registers were being introduced with a facility for collecting information on operator performance.

A second impact is through the increased possibility for collecting centralised information in computerised personnel records. The possibility of linking together information on work performance, discipline, time-keeping, together with information on wider questions such as trade union activities, is one that raises real problems of personal privacy. The chapter "Information Technology and Society" shows that the changes in individual privacy and freedom at work are linked to the wider problems of privacy and freedom of action in a computerised society.

The ergonomic effects of visual display units

Visual display units (VDUs), also called cathode ray tubes (CRTs), with keyboards have become the standard means of communication between people and computerised equipment, taking over from earlier technological means of communication such as punched cards and tape. This has led to the removal of a number of earlier computer-related jobs in data preparation. VDUs have also brought certain health and safety problems for the operators.

The use of VDUs can lead to particular problems of eye-strain, stress, fatigue, back pains, headaches and skin inflammation. An international trade union survey of the use of VDUs in ten countries found that the problems arose almost universally and in addition a number of unions agreed that monotony and social isolation at work increased with the use of VDUs and could have effects on health and well-being. The exact effects depend on a complex range of factors covering both the equipment and the work area. With regard to the equipment, relevant factors are: screen brightness; colour of display; character definition; size and space; "flicker" rates; tube implosion/explosion; radiation; heat generation; noise; cabinet design; maintenance; and keyboard design. With regard to the work areas, relevant factors are: lights and glare; paperwork; position of screen; keyboard and paperwork; space and overcrowding; posture. Other important ergonomic factors are the opportunity for regular breaks away from the equipment so as to allow human contact and rest. Without this the operator of a VDU can spend a whole working life looking at and interacting with a picture of reality rather than with reality itself. This can lead to very real social and psychological problems.

The importance of the new ergonomic issues thrown up by the use of VDUs had led trade unions throughout Western Europe to carry out studies and set out guidelines for operating with VDUs. A number of options exist concerning the overall impact of a technology, and the ergonomic problems of a particular process can be minimised and the opportunities maximised if the intention is to produce a healthy work environment rather than simply to increase productivity at any cost.

In general the requirements of a safe use of VDUs are that: neither a natural nor artificial light should cause glare on the screen; the characters on the screen should be easily readable; noise from the equipment should be kept to a minimum; the design of keyboards, desks and chairs should conform to ergonomic standards so as to avoid fatigue; VDUs should be positioned so as to allow social contact, but avoid overcrowding; operators should be given regular eye tests; operators should be given regular breaks away from VDUs. Research in some countries has also been directed at producing more ergonomically satisfactory VDUs through changing flicker speeds or developing new forms of displays.

Polarisation of Jobs

The conflicting effects of microelectronics seen in the simultaneous upgrading and downgrading of job quality and skills gives support to the view that it may be being used to further polarise and separate different groups within the labour markets of industrialised countries. Such a process of segmentation and polarisation of jobs in labour markets is not a new phenomenon nor is it solely the result of the introduction of microelectronics or even technological factors in general.

Analyses of labour markets in a number of industrialised countries show that labour markets are broken up into segments which together form a hierarchy of skills, pay, security, status and power. The segments at the top of the hierarchy include professional and managerial jobs but merge into the skilled trades and crafts in industry. These jobs require a high level of education and training whether obtained in schools and colleges or through experience at work. The jobs attract relatively high status, high pay and a general high rating of the job satisfaction, including for example freedom in carrying out work, career possibilities and at least until the onset of the current recession, high security. Lower down the hierarchy are what might be called semi-skilled jobs. These included the majority of lower grade clerical and administrative workers. The majority of semi-skilled jobs require a fairly basic level of educational achievement and the sector is marked by a fairly high degree of internal mobility. The degree of job satisfaction is generally low, although it varies considerably between different semi-skilled occupations.

At the bottom of the hierarchy are the unskilled and unqualified jobs with little educational or training requirements. Such jobs generally have low status, low pay, few career opportunities and a generally low level of job satisfaction. Work is generally highly supervised and controlled and there is normally a high rate of turnover. Attitude surveys suggest that unskilled workers attach greatest importance to job features such as security, physical conditions and earnings.

Much of the discussion of the segmentation of the labour market goes back to the industrial revolution of the nineteenth century in Western Europe. Through the division of labour, the simplification of a large number of tasks, and mechanisation based upon steam power, industrialisation and mass production took place. Technological change was undoubtedly one stimulating force behind the industrial revolution and it closely interacted with organisational and economic developments.

The economic rationale for the division of labour was perhaps first set out in Adam Smith's *The Wealth of Nations* published in 1776. This was further developed in Charles Babbage's *On the Economy of Machinery and Manufacturers* published in 1832, and developed perhaps to its logical conclusions in Frederick Taylor's *Scientific Management* published in 1911.

The economic advantages of the development of industrial society, based upon specialisation and technological innovation, has been the opening up of mass consumption and generally increasing living standards. Indeed in some countries at certain stages the workers have gone along with Taylorism in return for increased incomes. However, the social costs have been the intensification of the segmentation of the labour market with the creation of a larger number of standardised, simplified and repetitive jobs. The use that is being made of microelectronics to further polarise jobs is therefore not a new development, nor is it inevitable. It is rather a reflection of the ends to which the technology is being put. This has been summed up by the American computer scientist, J. Weizenbaum, in the statement that " . . . the computer, as presently used by the technological elite, is not the cause of anything. It is rather an instrument pressed into the service of rationalising, supporting, and sustaining the most conservative, indeed, reactionary, ideological components of the current Zeitgeist".

Choices do exist with regard to the uses to which microelectronics are put. Those choices will ultimately determine the overall impact of the technology upon working people.

The experiences in practice of implementing alternative systems based upon microelectronics have shown the positive opportunities that do exist for optimising other factors in addition to productivity in technological change. If microelectronics is introducted solely to increase productivity, then it will tend to lead to a polarisation of skills. For example, the introduction of computer numerically controlled (CNC) machine tools had the effect of deskilling the machinist's job when used to economise on labour, but could also lead to an enhancement of skills if introduced with a flexible form of work organisation. The same choices exist with the introduction of word processors in areas of clerical work. The new technology could be used to increase work intensity or to reduce the amount of time a secretary need spend in typing.

Technological Alternatives

A study of West European experiences of job redesign published by the European Trade Union Institute has shown that concrete examples exist of job redesign improving working conditions. For example, job enrichment which gives the right balance of administrative and typing functions, can give positive ergonomic results. As seen, regular breaks away from visual display units (VDUs) are essential for health reasons and a rotation of tasks can ensure that a secretary spends only a limited period of work at a VDU. Trade unions have pressed strongly in most West European countries to ensure that secretarial jobs are designed to require only a limited period of time operating word processors. This has resulted in a number of collective agreements being reached which cover this aspect in different countries. For example, a July 1980 agreement in the Federal Republic of Germany stipulates that VDU operators in the bank sector should spend no more than four hours per day in front of machines. The agreement also stipulates a certain time for hourly breaks, and regular medical checks. When coupled with the observation of ergonomic standards such as those set out earlier in this chapter, job rotation can significantly reduce the hazards of VDU operation.

The application of microelectronics in products and processes has also provided a stimulus for the decentralisation of work organisation towards small independent production units which are centrally co-ordinated. On the process side, the introduction of computer systems allowing distributed data processing facilities, but linked to a central computer, can facilitate such forms of work organisation. As already noted, the introduction of new technologies can lead to a breakup of line production of small independent production units.

The introduction of microelectronics in products may also provide a stimulus for changes in work organisation. In one Italian business machinery company, a series of reforms in the system of work organisation were introduced in the 1970s, as a result of both union pressure to decrease the monotony of certain jobs and the management desire to achieve a more flexible system of production in the light of the switchover to electronic-based products from mechanical products. The result was that integrated assembly units were introduced to replace line assembly, with the individual unit being responsible for a specified product or unit assembly. This work structure was greatly facilitated by the modular nature of the production process with electronic products.

The opportunities for decentralisation of production systems given by microelectronics can, however, improve or worsen working conditions. As pointed out earlier, the possibility of increased homeworking can lead to an increase in social isolation of the worker. The possibilities given for the creation of autonomous work groups can also be used to undermine trade union organisations and fragment the work force if not accompanied by facilities for social contact and meetings between work groups. Again the impact of the technology will reflect the social choices made in its use rather than any technical imperative.

One of the problems associated with microelectronic-based technology is that it may be used to obscure basic social and economic choices, with technical complexities mystifying those non-specialists affected. In addition, microelectronic-based, computerised equipment is introduced in stages so that at the point of introduction of an individual piece of equipment, the overall effect upon jobs is obscured. However, through the linking together of individual pieces of equipment, the impact of the overall system may be considerably greater than the parts. For example, the introduction of electronic point-of-sale terminals in retail stores may be part of a general programme of computerisation covering computerised warehousing, bar-coding and stock control, with implications upon employment much greater than might be apparent at the point of their initial introduction.

The Industrial Relations Implications

Experience has shown, therefore, that microelectronics can be used to achieve a less polarised structure of industrial society or it can be used to increase centralisation and polarisation. The choice is a social and political one, and is not technological.

As has already been shown in this chapter, if jobs are redesigned so as to allow responsibility, variety, control for the worker, then microelectronics can be used to improve working conditions and upgrade skills. The importance of only spending a limited period of time in operating a computer terminal has been shown to be an important factor in determining job quality and when new technology is introduced, jobs should be designed with this in mind.

Improving working conditions may involve a conflict between the interests of workers and the interests of management. At certain points, however, the long-term interests may allow short-term conflicts to be negotiated, resolved and a "trade-off" achieved. This is after all the logic of industrial relations based upon collective bargaining. "Change by agreement" is the only viable long-term principle for effectively managing change.

The attainment of the long-term optimum balance between the maximisation of social and economic goals requires, therefore, participation of all those affected by the change process. For working people, who are the main direct users of new technology, this participation will involve considerable changes for their own representative organisations, their trade unions.

In the face of past periods of change, many trade unions have developed defensive responses to change by trying to protect the interests of their members at a stage when decisions could not be influenced. Important as such action may be when all else has failed, it still only amounts to a fire-fighting role. To ensure that technological alternatives are not wasted, trade unions must obtain information on changes at a stage in the process when real technical and social choices can be influenced and to shift their policies to being active rather than reactive.

This is rather a new path for trade unions in many countries requiring changes in many traditional areas of activity. In some countries new bargaining levels have had to be introduced, particularly at the company level. Demarcations between industrial or craft unions frequently become blurred as new technology transforms the nature of jobs and leads to transfers in jobs between sectors: this requires changes in union structure. The importance of a close liaison between the union negotiators and work groups also becomes increasingly important. Furthermore, microelectronics increases the importance of trade union education, training and liaison with academic specialists and others not directly part of trade union structures.

On the management side, some of the implications of microelectronics have already been outlined in the chapter "The Impact on the Enterprise". Some traditional attitudes towards management prerogatives and the scope of collective bargaining will have to adapt. If societies are to become increasingly democratic, then it is inconceivable that high living standards, political awareness and democracy in social life in general can co-exist with an increasingly polarised and Taylorist work structure.

In several industrialised countries, developments along these lines are beginning to take place. For example, in Norway a national level procedure agreement was signed in 1975 between the national trade union and employers' organisations, covering the introduction of computer-based systems. The agreement provided for the provision of information to the unions by management concerning proposed changes as early as possible, before decisions were made, and it also stipulated that both management and unions should provide information and involve the workforce to the greatest extent possible in the planned changes. The workforce was given the right to elect specialised representatives to deal with technology questions and to be fully trained with computer-related skills, and these came to be known as "data stewards".

In a number of other countries important sectoral agreements have been concluded, particularly in those sectors most directly affected by microelectronics. In most West European countries, for example, agreements have been concluded in the newspaper industries covering the introduction of computerised typesetting. These have typically covered questions such as health and safety, manning levels, retraining and mobility, downgrading limits and income guarantees for the displaced craftsmen. In some West European countries, sectoral agreements have also been concluded covering the introduction of new technology in the post and telecommunications sectors (e.g. the United Kingdom and The Netherlands), the metalworking industries (e.g. parts of the Federal Republic of Germany and Italy).

Despite such national framework agreements and sectoral agreements, it has been at the company or organisational level where perhaps the most important decisions are taken concerning the application of microelectronics and strategic planning. In a number of countries, trade unions have developed company level organisation to allow negotiation of technological issues. In the Federal Republic of Germany the works councils have been one of the main instruments through which the trade unions have sought to obtain information and influence over the introduction of new technology at company level. They have the legal right of co-determination, consultation or information on general issues of company policy which includes technological change, although not of negotiation in terms of the conclusion of collective agreements.

In the United Kingdom a number of technology agreements have been signed at company level. The bulk of these have covered the procedures for consultation and early warning of the introduction of new technology, and the relationship of existing trade union machinery or procedures to those created to handle new technology. They have gone on to lay down specifications for job security, limits on downgrading and the retraining requirements. Agreements typically also cover the ergonomic requirements for operating computer terminals and visual display units, and include the design of the job, including work organisation. The final area that most agreements also cover is the level of working time and also the pay implications of the introduction of new technology.

The development of effective negotiation on the introduction of new technology has re-emphasised the importance of ensuring a close relationship between the elected trade union representatives and the membership, and indeed for developing a full education programme for the mass membership with regard to the technology. For example, in Norway, from the early 1970s, the national computing centre and the unions in the metal industry co-operated in developing educational courses for union members, and this resulted in the publication of two textbooks for trade unionists on data processing, planning and control. This co-operation between academics and unions was subsequently repeated in other sectors, and in other countries, particularly in Denmark, the Federal Republic of Germany and Sweden. One of the most important conclusions of these experiences was the need to encourage rank and file trade unionists to fully analyse their own objectives with regard to technological change.

The Swedish unions have also developed the use of "wage-earner" consultants. In some cases, there are examples of projects being developed by the unions with the help of systems consultants whose fees are paid by the employers. The unions have also realised the importance of having access to research and development (R & D) resources independently of the employers, and this has led them to attempt to extend their influence over the allocation of the part of the national R & D budget which is directed towards applied industrial R & D.

The general education structures of industrialised countries will also need to adapt to changed circumstances, not simply to provide more computer experts to meet the increasing demands of computer manufacturers, but to ensure a wider distribution of computer-related skills. This means the demystification of computing in the general education system and at the same time specialists must be made aware of the social and organisational consequence of the diffusion of computerised systems.

Finally, governments must act to provide a macroeconomic environment, against which technological change can be effectively managed. Collective bargaining will only be able to deal with problems of technological change against an overall background of growing employment, living standards and the satisfaction of workers' social aspirations. These problems are referred to in the next chapter on "Microelectronics and Macroeconomics". The challenge of the 1980s must therefore be to effectively tackle the problem of unemployment through a combination of means including the reduction of working time, a co-ordinated action programme for economic recovery and the direction of growth towards employment creation. Unless we do this, then the potential offered by new technology will be wasted in higher unemployment. Against this background, workers will fight to have security *against* change instead of security *in* change.

6

Microelectronics and Macroeconomics

GÜNTER FRIEDRICHS

The previous two chapters have dealt with the effects of microelectronics at the level of companies, plants and working places. Now the effects on macroeconomics, that is to say on the overall economy, are to be investigated.

Technology has always played an important role in the industrial process. However, technology has never been an autonomous force. It has always been a result of human activities and a tool of human beings. So far, technology is neither good nor bad. It just depends in what way it is used.

In the long run improved technology has been one of the main instruments for the development of wealth in the industrialised countries. Technology has been used to increase the productivity of human work or to decrease labour input per unit produced.

However, new technologies are not only more efficient. At the same time they can be destructive, by replacing existing technologies and human work. They may destroy jobs, qualifications, machines, plants and even enterprises. In the long run new technologies tend to create new structures and jobs. In the short run they very often create economic and social difficulties. The extent of the adjustment processes needed to handle these problems differs. It very much depends on the innovative power of the new technology. In the case of microelectronics the innovative power seems to be so big that even the international distribution of labour will be affected.

Microelectronics—A Key Technology

There are several points that make microelectronics the key technology of our decade:

(1) The possibilities for application are so many that one can say that all parts of the economy and the society will be affected in some way.

(2) The drop in the prices for important hardware is so rapid that high-speed diffusion can be expected.

(3) Automation in the past was restricted to mass production, because this technology did not offer the type of flexibility necessary to adjust to changing market conditions. Production thus still depended on the availability of experienced human beings. In the future microelectronic control and steering devices will offer a high degree of flexibility for automatic production in small- and medium-sized batches.

(4) In the past technical change affected production much more than offices. In the future microelectronics will play an important role not only in production but also in private and public administrations and within the service sector.

So far, microelectronics seems to be a basic innovation, comparable with steel, electricity or automobiles.

The foregoing chapters already showed many actual or future applications of microelectronics. For this reason this chapter concentrates on major applications where larger numbers of persons will be affected. Since at this moment everything which can be said is more or less speculative, attention will be restricted to foreseeable developments up to the end of the 1980s, without mentioning exact dates. First of all areas of industrial production are to be investigated.

Production

Many people believe that the *production* of new labour-saving equipment creates new jobs. Some of them are prepared to accept the idea that there could be a negative effect at the level of the users. In the case of microelectronics, statistical data on the production of office and data-processing machines within the Federal Republic of Germany are available.

We will compare them with the average 1970-1979 figures from thirty-four sectors of mining and manufacturing industries and with plastic manufacturing. Plastic manufacturing is included because it increased most in terms of employment and production in spite of high productivity increases. Plastics itself is a key technology that affects many different areas of the economy, however, its importance is much smaller than that of microelectronics.

Indicators of Technical Change in the Manufacturing and Mining Industries, Office and Data Processing Machines, and Plastic Manufacturing in the Federal Republic of Germany, 1970-1979

	Production in per cent	Employed persons in per cent	Volume of employment in per cent	Productivity per hour in per cent	Potential capital per output ratio in per cent
Mining and manufacturing industries	+ 21.4	—14.4	— 22.6	+ 56.8	+ 2.8
Office and data processing machines*	+ 74.5	— 16.0	— 19.9	+ 117.9	+ 1.0
Plastic manufacturing	+ 98.6	+ 26.6	+ 16.0	+ 71.2	+ 8.8

Source: DIW, 1981
*These figures cover all areas of the production and sales of office and data processing machines, including software personnel of the producers, however, excluding software houses and software personnel of the users.

Out of a total of thirty-four different sectors of the mining and manufacturing industries, the manufacturers of office and data processing machines were able to realise the second highest production increase between 1970 and 1979. But in spite of a production increase of 75 per cent, in 1979 employment was still 16 per cent below 1970, because of an increase in productivity (production per hour worked) of 118 per cent. And it is the increase of production per hour that represents the labour-saving effect of improved technology. The employment figures in computer production are better: employment is in fact increasing considerably. However, West German statistics, in accordance with ECE (Economic Commission for Europe) standards, combine the figures of the data processing and office machine industries. The macroeconomic weight of the office and data processing machines combined is very limited. In 1979 their percentage in value added was 1 per cent and in employment 0.9 per cent of the total of the manufacturing and mining industry.

The situation of plastic manufacturing is somewhat different. As already mentioned, plastics also belong to the leading new technologies. Between 1970 and 1979 this sector reached the highest growth rate of production of all and was one of the four sectors which had a higher rate of employment in 1979 than in 1970. The increase was 26.6 per cent. This industry is happy to achieve growth rates of production that are able to overcompensate productivity increases. In plastics this may well continue because the industry can expect special benefits from the introduction of microelectronics, since a combination of plastic material with microelectronic instrumentation has very often proved successful. But we have to keep in mind that each gain for plastic material is at the same time a loss for conventional materials and for those producing them.

One of the areas which has had the most spectacular labour savings is the replacement of mechanics by microelectronics, as happened in the watch-making industry or with the production of telephones, telex-writers, taximeters, sewing machines, and many others, since the number of individual parts that had to be produced and assembled could be greatly reduced. The result was saving in manpower between 40 and 50 per cent as was shown in the foregoing chapters.

Another application of microelectronics in production is in machine tool building, especially the numerically controlled (NC) machine tool. Until about the spring of 1978, diffusion of the relatively old and very well-known numerically controlled machine was very slow. The reason was that expenditure for measuring, controlling, and steering devices accounted for about 50 per cent of the total cost of a NC machine. In the past these functions were done by very expensive conventional electronics. Now conventional electronics can be substituted by cheap microelectronics. And indeed at present the first generation of relatively cheap computer numerically controlled (CNC) machinery is on the market. The costs for steering devices dropped from 50 to 10 per cent. Now we can afford to connect CNC machinery directly with computers and to integrate them into direct numerically controlled (DNC) systems. Under these conditions we can expect a rather rapid diffusion of this technology. On average, one CNC machine tool saves two jobs.

Another area where the application of microelectronics has similar effects is process control. The continuous process industries will be affected the most (oil, chemical, cement, paper, etc.). It will be possible to substitute jobs which have remained from previous steps in the automation process for measurement and control by small computers. However, there are also many opportunities for process control in the metal trades. This is especially true for computer aided manufacturing systems (CAM) including computer aided engineering (CAE), computer aided design (CAD), computer aided planning (CAP) and computer aided quality assurance (CAQ). All these systems are well known and some are very far advanced in development. When they are offered at reasonable prices, they will be adopted very soon. An interesting example is a West German foundry employing 150 persons. There the changeover from manual drawing and construction to computer aided design has been completed. Formerly they needed two "Ing.-grad." (technical college graduates in engineering) and 15 technical draughtsmen. Today they need just one "Dipl.-Ing." (MA in engineering), one terminal, and one plotter. Another "Ing.-grad." is necessary only for situations when the "Dipl.-Ing." is not available.

A last example that should be mentioned here is the industrial robot. Probably 1979 can be called the year of the big breakthrough in automatic handling equipment. Many car manufacturers installed automatic welding assembly lines using robots. The present generation of industrial robots is already equipped with microelectronic steering and control devices which can be used for many handling tasks within different types of industry. However, they cannot achieve the level of precision which is necessary for most assembling jobs. Again microelectronics is being used in the development of sensors for the next generation of robots which is already available as pilot installations. These so-called "intelligent" robots will be soon available.

They are able to take over a big proportion of all handling and assembling jobs. If they do so, they will destroy most of the existing assembling lines. This might be good for the quality of work, however, it will increase employment problems especially for women. On average, the present generation of robots replaces four workers and creates one new job of taking care of the robots.

After this discussion of examples of industrial production, the effects of microelectronics on the office are to be examined.

Administration and Offices

The most important field of application of microelectronics here is the use of computers. Therefore, data processing will increase in offices as it becomes cheaper and more available also for small business. Since many private and public administrations have become very familiar with data processing, they will seize their chance to use it. Not only data processing centres but many workplaces as well will be equipped with or connected to decentralised microcomputers. The number of visual display units (VDUs) for data input, controlling operations, interactive systems is increasing considerably. This opens the way for the computer to enter the small business, something which was excluded until now because computers were too complex, difficult and expensive.

In addition, a very exciting development within our decade will be automatic typewriters and programmed text processing. The main threat seems to be for typists and clerks.

The possible impact of these developments can be seen from an investigation which looked at the office of 1990 and concluded that a high percentage of normal work can be formalised or automated. The investigation covered 2.7 million office jobs: 42 per cent could be formalised and between 25 and 30 per cent could be automated. The potentials for formalising and automating differed in the areas investigated. In public administration big saving potentials were recognised, since about 75 per cent of all jobs could be formalised and 38 per cent could be automated. Within the private sector possible savings of between 25 and 38 per cent were detected. The investigators believe that the office automation potential will be realised by 1990. If this is true, a great percentage of typists and clerks will be in difficulties.

Since administrative and office work was considered to be of low productivity in the past, many economists expected this area to be an option for the employment of additional people who could not be employed any more in production due to technical change. However, this is not the case, since technology is penetrating this sector as well. The word processor has been well known since the beginning of the 1960s, but its price was too high for wide adoption until recently. Now there are many different systems competing with each other and users are still trying to find out which serve their needs best.

In 1980 the ratio between prices for standard typing automata (including memory) and prices for electric typewriters became really attractive and the breakthrough started.

The automatic typewriter has two important potential cost advantages. One is the possible replacement of mechanical equipment by microelectronics; the other is the enlargement of storage capacities. The first change cuts down quite much on cost and the second one will not increase the prices considerably.

In March 1981 the first Teletex system in the world was started by the West German Post Office. Automatic and memory-equipped typewriters are now able to communicate with each other at a speed of 300 signs/sec (2400 bits/sec) by using an existing network. The quality is exactly the same as that of a typewriter and the machines are able to send and receive at the same moment, by using two different memories. This will, in time, replace the telex system. Also in 1981 a U.S. company offered a network based on a special coaxial cable and able to link up terminals, automatic typewriters, printers, copiers and other electronic parts of the automated office.

An even more advanced technology would be word processing without any paper or at least using only computer-printed paper. All files could be made available by either visual display units or printouts. Special memories could store text and data for recording. That development could be called the paperless office.

Even in the face of these developments, some experts do not expect large dismissals due to office automation. They claim that the rate of labour turnover in offices will continue to be very high. However, this would still mean no job openings, a decline in office jobs, and increased unemployment.

The Capital-saving Effect

Until now the labour-saving effect of microelectronics was considered to be the ability to produce a bigger or at least the same amount of production with less labour input (the labour productivity concept). But what about the effect on capital or respectively the productivity of capital (the capital per output ratio)? In the past, technical change was connected with a slow but permanent increase of the capital per output ratio. Only a few industries such as the chemical industry and oil refineries became more effective by saving capital instead of manpower. Both produced a very clear type of capital-saving technical change, in spite of expensive anti-pollution regulations.

The capital per output ratio should not be mistaken for the intensity of capital, that is, the amount invested per employed worker. This indicator will increase as long as technical change has a direct labour-saving effect.

However, the capital per output ratio represents the capital cost per unit produced. A decrease in this indicator would have a tremendous impact. In the case of microelectronics this seems to be possible if we make a distinction between miniaturisation and associated cost reductions and normal products with incorporated microelectronics.

In the case of miniaturisation and associated cost reductions like data processing, word processing, numerically controlled machine tools, there are cost reductions already and these will be extreme within the next years.

This is the reason why very successful types of applied microelectronics, like computers for many different purposes, computer numerically controlled (CNC) machine tools, word processors, copiers, or from the very beginning pocket calculators, are offered as low-cost systems. Possibly robots and computer aided systems will follow in the coming years.

This is different in some cases where microelectronics is incorporated in existing products. Normal incorporation increases product quality, however, it should not increase prices. Possible decreases depend on the amount of incorporated electronics in relationship to the total value of the product. But only in cases of miniaturisation and associated cost reductions, is the capital per output ratio affected. Then a destruction of working places is involved, because the depth of production is being reduced.

All cases of capital saving have at the same time two labour-saving aspects. The producer enjoys the lower production costs, and the user enjoys the same thing but from his point of view. The overall economic result is a first labour-saving effect through producing investment goods and a second labour-saving effect in manufacturing end products.

A general development in favour of capital-saving equipment is very unlikely. However, a development like this in several important areas of the economy is probable. It does not make much difference if low-cost systems are so successful that they reverse the present capital per output ratio of the total economy or only of some important sectors. Even in the latter case it would result in a slowdown of the present increase of capital cost per unit. This will affect negatively the chances of creating new jobs through the production of investment goods.

The effect on employment through the need for additional software people will be discussed later on. At this stage it should be noted that the cost of software has to be added to the capital costs. However, in spite of very urgent software needs and in spite of an obvious time lag between the development of hard- and software, the costs for the latter will not exceed the savings from the first.

How About a Post-industrial Society?

How will microelectronics affect the future structure of the economy? C. Clark and J. Foruastié expected that the so-called tertiary sector of the economy or, in other words, the less productive private and public administrations and offices as well as the less productive private and public services, would be able to compensate job losses in the highly productive areas of industrial and agricultural production. Since this development has been confirmed statistically, we are thinking today in terms of a post-industrial society.

In industrialised economies the number of people working in agriculture is dropping and will continue to do so. The far bigger decrease in the number of persons working in industrial production will continue and there will be very limited chances to transfer them to the administrative or office spheres of industrial companies.

For example in 1970 the mining and manufacturing industry of the Federal Republic of Germany still needed nearly 8.9 million people. In 1979 this number had dropped to 7.6 million, that is, −14 per cent. Moreover, the volume of employment (number of employed persons multiplied by hours worked) showed an even greater decrease—23 per cent—in spite of the fact that production in the same period increased by 21 per cent. In other words, a production increase of 21 per cent was accompanied by a drop in the volume of employment of 23 per cent. This development shows that the growth rate of the West German mining and manufacturing industry in the 1970s was far too low to compensate for the high productivity increase (production per man hour) of 57 per cent.

For this reason we need to determine whether in future the public or private service sector can absorb persons displaced from agriculture or industrial production or office work, as it did until the end of the 1970s. Therefore, we need to check the statistical trends in the service sector. First of all those sectors must be listed where decreasing employment can be expected.

Parts of the Service Sector with Decreasing Employment

Transport: (1) information, (2) railways, (3) shipping. All these sectors already have either stagnating or decreasing employment. This development will continue due to the effect of microelectronic applications.

Distribution: data processing, text processing and especially point of sale terminals as well as concentration make certain that the number of employed persons will continue to decrease.

Banks and insurance companies: data and text processing combined with integrated accounting terminals (on-line banking) will lead to a reduction of employed persons, or at least to a stagnation of employment in spite of a strong increase of business volume. In banking the reduction of jobs started in 1981 and will continue in the coming years.

Public administration and offices: data processing, text processing, copying services and some other labour-saving technologies will prevent the number of employees from increasing.

The social security legislation of the public services has very effective controls against firing, downgrading, etc. However, this will not prevent the reduction of staff. If necessary, only a few of those who are leaving because of age, marriage, children or any other reason will be replaced. The number of employees in public administration will decrease more slowly than in other sectors, but it will decrease. However, this will be different in some specific public services.

Parts of the Service Sector with Increasing Employment

There are some areas within the service sector where we can expect and where we need employment increases. This will be partly public jobs or jobs influenced by public administration. The spheres are:

— education, including further education;
— research and development;
— social services in the widest sense, including health, advising, counselling, rehabilitation, etc.;
— hotels and restaurants;
— transport by truck;
— transport by air.

According to the hypothesis of this paper, namely, that microelectronics will play the leading technical role within the next decade, there seems to be no room for an industrial or a post-industrial society. Both sectors, industry as a whole and public and private services, will lose employment possibilities which now exist. And agriculture will continue to replace workers and farmers through the effects of the so-called "green revolution". There will be expansion in production as well as in offices of industrial companies and in services. However, there will be no corresponding increase in employment. Employment in certain sectors of industry as well as in the services may expand, but the increases will not be big enough to take over all those who have employment difficulties. In certain sectors, especially in those which incorporate microelectronics in their products, some companies will achieve growth rates big enough to hire additional personnel. At the same time, other companies will be able to increase sales but will not be able to keep their labour force stable. And again, since this new technology offers ever new opportunities, there will be many companies which will fail by not taking into account the risks of innovation and others which will fail because they do not see the need for innovation.

As a result of this development, many highly developed countries with already high rates of unemployment will not be able to reduce these rates and indeed will be in permanent danger of seeing unemployment rates rise even higher. The persons who will be affected most are women.

International Trade

Microelectronics will also affect international trade and the international division of labour. Since the general economic conditions are worsening, one should not be surprised at increasing international competition. At least the highly developed countries will try to improve their situation by increasing exports. This is normal and acceptable, however, it also means that they are in fact trying to solve their problems by exporting them to others. It is also understandable under these conditions that some countries will try to defend themselves by raising their import barriers. The result could be a new area of protectionism in spite of the fact that the world economy in the past developed best in periods with low degrees of protectionism. That these periods of supposedly satisfactory development were not satisfactory at all for many developing countries is another part of the same story.

However, there is a very remarkable development in connection with microelectronics. Especially the highly industrialised countries have invented sophisticated methods very close to protectionism, or not to say mercantilism, by not raising tariffs or setting up other import restrictions, but by financing and promoting the research and development (R & D) activities of their industries. This happened in the United States by way of NASA and military budgets, in Japan, in France, in the Federal Republic of Germany and in the United Kingdom through government programmes.

Within the group of the highly developed countries, the United States is still leading in microelectronics, followed by Japan which possibly has a chance to take over. The leading companies in Europe are not in a leading position. Therefore, they are trying to find partners in the United States and Japan, and they are taking over companies in those countries if they can get them. Certainly governments do not subsidise transactions like this directly. However, the indirect effects of subsidising R & D make these transactions much easier.

At present the race is concentrating on the production of memory chips within the category of very-large-scale integration (VLSI) and in future will focus on high-speed integration. The industrialised nations believe that they need direct access to the production of highly integrated and sophisticated chips. Otherwise they fear to get dependent on the successful producers. This is the reason why governments have developed new methods of helping their national industries. And from their point of view they have no other chance. They should not be surprised, however, if other nations and especially the less developed countries find a chance of defence only by increasing import barriers in the classical way.

On the other hand, it is still an open question whether access to the chips is really so important. There are at least two countries that can offer chips in any desired quantity and quality. But about 10 per cent of the value of the average product which incorporates microelectronics is for the chips and 90 per cent for the rest of the product. Some people argue that the chips are not so important because one can buy them, but what is the most important is the nation's capacity to produce the other 90 per cent. The defenders of this strategy want all governmental help to be concentrated on the application of microelectronics, which is yet another, different, measure of protectionism. However, it is difficult but still rather easy to find international agreements about tariffs or other import limitations in comparison with international agreements about R & D activities. On the other hand many industrialists defend their activities in favour of self-produced and highly integrated semiconductors with the argument that otherwise they need to leave their specific know-how to the producers of chips. For this reason they may just concentrate on the design of specialised chips that need to be produced by a normal producer.

At this time nobody knows what the final outcome of the present R & D competition among the highly industrialised nations will be. But the motivations of the participants in the race are quite obvious. All fear to lose their important comparative advantages in the future. This is the reason why they are prepared to introduce policies with protective effects that are not yet identified by the less developed countries.

The introduction of low-cost systems with a high degree of automation may have a reverse effect on the present international distribution of labour. Many high-wage countries tried to create production facilities in low-wage countries. This did not work in countries without any industrial tradition. But it works in several of the more advanced developing countries. However, the application of highly automated low-cost systems offers new opportunities for the return of production to high-wage countries. Especially the combination of low-cost systems with highly qualified labour offers the possibility for a kind of flexibility that cannot be provided by developing countries.

There are only two results of the present R & D competition which we can expect definitely. (1) The so-called "rich" countries will get richer and the "poor" ones will get poorer; not necessarily in absolute terms but in relative ones. (2) The present rank order within the group of industrialised countries will change considerably. Some will gain and others will lose, however, not necessarily only in relative terms but also in absolute ones.

The same will happen to the developing countries. The low-cost systems will be available to them and the cheap prices seem to be helpful. But these countries will have big difficulties with software. The software available in industrialised countries will be different from what is needed in most developing countries. A few of these countries will be in a better position: those with large numbers of unemployed graduates. However, the incapability to produce sophisticated hardware, the difficulties to produce appropriate software and the dangers of a brain drain to the industrialised countries will make it extremely difficult for most developing countries to keep up. Some of those countries that are used today by highly developed countries for certain production activities in the field of semiconductors may have a chance if they behave extremely intelligently. But what can be expected in general is a complete change in the international division of labour to the disadvantage of most developing countries. The chances of developing countries will be restricted to two strategies: (1) concentration on relatively simple incorporation of microelectronics in mass products and (2) the production of software related to the specific needs of the developing countries.

Microelectronics will anyhow contribute to making employment problems of the developing countries even much more difficult than they are today.

Centralisation versus Decentralisation

Microelectronics will also affect location and concentration. In the past, technical change contributed considerably to the centralisation of economic activities. This might be changed to some extent by microelectronics. The coming new forms of communication allow decentralised production and decentralised administration. This certainly does not mean decreased power for the multinational or big corporations. It does mean the possibility of becoming much more independent from location. Administration, planning, steering of production, and production itself can all have different locations. This will allow settlement in regions that are outside the classic big economic centres. This opens options for improving the economic conditions of both the less developed regions as well as the big activity centres.

These new forms of communication will certainly strengthen the power of big corporations and multinationals because they will benefit from much more flexible systems of steering and control. However, small- and medium-sized business can also profit from microelectronics. Certain types of production machinery as well as data and text processing, which were relatively expensive yesterday and therefore in a certain sense exclusive, will be and partly are already available to smaller companies and to small business. Microelectronics will perhaps repeat what once happened with the introduction of the electric engine: the survival of small- and medium-sized business in a world where the big ones continue to get bigger.

The Employment Problem

The most difficult problem will be to attain and retain full employment. The introduction of microelectronics will reduce employment possibilities in production, offices and services under present conditions. In the past, at least in many periods, the labour-saving effect of technical change could be compensated by relatively high rates of economic growth. This was especially true in the 1960s. This did not help the individuals or companies affected but did help the economy as a whole. Ever since the beginning of the seventies all industrialised countries have had difficulties in reaching satisfactory growth rates. And the prospects for the eighties look even worse.

The first question that should be asked is what contribution microelectronics can be expected to make to economic growth. A first answer was given by the figures of the West German office and data processing machines industry referred to earlier in this chapter. They showed an enormous increase of output and at the same time a decrease in the volume of employment. The reasons were the extreme increase in production per hour (productivity) and the considerable narrowing of the depth of production.

The next question to be asked is whether a given product which contains microelectronics serves investment or consumption. Investment goods are used for production purposes. Therefore, they are bought only if their efficiency is higher than that of the goods they are replacing. So we can expect that all investment goods using microelectronics will have a high labour-saving effect. This is certainly true for office and data processing machines which are not only produced with high productivity increases but will also create themselves additional productivity effects. Their users use them for labour-saving process innovations. And this effect is also true of many other different investment goods incorporating microelectronics. Under these conditions the production of investment goods will create new jobs only in times with high rates of economic growth and/or very highly increasing exports. And this is something nobody expects in the years ahead.

The situation is different if the product serves final consumption. In this case we have to distinguish between four different categories: (1) existing products are becoming redundant, (2) existing products can be produced with higher efficiency, (3) existing products have higher quality through additional functions, (4) completely new products are introduced which depend on the existence of microelectronics.

Looking at these four categories, it becomes quite obvious that jobs are definitely destroyed if existing goods become redundant. The same is true for existing products which can be produced with higher efficiency, with one exception: if the product is sold so cheaply as to stimulate demand. The producer may be able to overcompensate his manpower savings by increased production. Whether products with higher quality through additional functions provide more employment is an open question. Improved quality through additional functions because of microelectronics may still be labour saving. However, such products offer a chance for increased sales and therefore, at least theoretically, the possibility of additional jobs. It certainly depends on how the consumers react to the improved quality. In some cases even very high quality improvement (for example colour TV) needed a long time to be adopted by the majority of consumers. But it would be very optimistic to expect an extremely high increase of durables just because of quality improvement based on microelectronics in the highly developed countries.

From the employment point of view only products with improved quality and with fast diffusion rates and moreover products which are completely new are of vital importance. However, up to now the number of really new products, for example TV games, has been very limited. A typical example is the "home computer". This product is not presently accepted by families; however, it has reached big sales to small business. This is not astonishing. At the present stage, and this will continue for some time, microelectronics offers so many possibilities for saving labour and capital in investment goods that producers and users are primarily interested in this aspect. Since many well-known investment products can be produced cheaper and very often with improved quality, it is rather normal that everybody tries to make use of these chances. The innovation of a completely new consumption product involves a much higher risk than the cheaper production of known goods. An OECD study from September 1980 concluded: "... there is no evidence that information goods and services may in the near future become an important growth factor in final demand. Consumption of information goods and services is still playing a fairly minor role in the budget of the average household."

In the very long run, however, microelectronics offers both quality improvements that are so high that additional demand is created and completely new products that will definitely create new jobs. However, there will be a long time lag between the period in which the use of microelectronics is dominated by process innovation and the period in which product innovations start to dominate.

The problems are (1) what to do within the time lag, and (2) whether or not in the period of product innovation microelectronics will produce so much growth that we can have full employment under present conditions.

In spite of some exceptions, the innovation period for consumer products will not start before the next five to ten years. Even if this development reaches full speed, it must be questioned whether full employment will be reached again and can be maintained through many new products. And it is even more questionable whether governments can afford another period of five, ten or even more years with increasing unemployment.

It is not the new technology alone that increases unemployment. Demographic reasons, increasing prices for energy and materials, inflation, decreasing demand for durables, worsening developments of the terms of trade and insufficient rates of economic growth all contribute to increasing unemployment in the industrial countries.

Even in Japan, a developed country with very prosperous growth rates, a working group at the Research Institute of the Japan Management Association under the sponsorship of the Ministry of International Trade and Industry concluded in November 1979: "... In view of slower economic growth and the inevitable changes in industrial structure that are foreseen in the 1980s, Japan will sooner or later face the problem of interaction between technological innovation and employment. In particular, if this country chooses to emphasize knowledge-intensive industries in the 1980s, it will not be able to avoid employment problems arising from the extensive use of microelectronics.''

How About Software?

In one area microelectronics is already job creating. This is because of the need for additional software. More programmers are needed, more system analysts, and so on. They are all necessary to keep the systems running. These activities are job creating and they may compensate for some disadvantages resulting from the capital- and labour-saving effects of the hardware.

But we cannot assume that the production of software will absorb all those who lose their jobs through the introduction of new hardware. Neither can we assume that the individual production of software by individual users and parallel work of several companies for the same purpose will continue. Producers and consultants will offer standardised programme packages which only need to be adapted to the needs of individual companies. The adaptation will only create a limited number of new jobs.

On the other hand, in spite of the high increase of activities, existing computing centres will not increase but as a matter of fact they will decrease their staff, due to productivity increases and due to the more decentralised types of organisation. It must also be considered that EDP (electronic data processing) producers will try to integrate software elements into their hardware. If they are very successful, it could mean that hardware will become less compatible, which creates quite different problems. However, in terms of employment it must be said that any progress in software integration into hardware limits the chances of additional software employment. The above-mentioned Japanese study starting with 80,000 software engineers in 1975 expects between 282,000 (minimum), 449,000 (medium) and 796,000 (maximum) software engineers for 1985.

In other words, there will be new jobs for the production of software. But they will not be able to compensate for the total job losses. And these figures demonstrate a very important point of strategy. The Japanese seem to consider the availability of enough software engineers as the key to success!

Two competing studies about the economic and social effects of modern technology, sponsored by the government of the Federal Republic of Germany, came to very pessimistic results. One predicted that productivity increases in the 1980s will be higher than economic growth and therefore unemployment will rise. Another study forecast a rise to 1.6 million registered and 1.4 million hidden unemployed, i.e. 3 million jobseekers, in 1985.

Reduction of Working Time

Two suggestions can be made for fighting unemployment. The first is not very original. This is a better distribution of available work between those who would like to work. This can be done by cutting down working time. In this respect it is important that the total working time for each individual throughout his life needs to be reduced. Employers are fighting extremely hard against any move in the direction of a thirty-five-hour week. German employers, for example, are prepared to accept, and in many cases have already accepted, a six-week annual vacation which is to be implemented by 1982 and they are also prepared to pay 50 per cent extra money for the vacation period, since it is more expensive to be on vacation than at work. However, they do not want the forty-hour week touched. Since the thirty-five-hour week is by far not the only way of reducing working time, the unions need to think in different categories in order to achieve the same purpose. But there is no possibility of preventing a shorter working week in the long run. The arguments that are used today against the thirty-five-hour week are exactly the same that were used when the unions fought for the forty-eight-hour or the forty-hour week. In the future there will definitely be shorter working hours and this may create completely new conditions for working life and also for consumers demands.

On the other hand, earlier retirement and additional free and paid time for further education could also have a positive labour-market effect. The expansion of further education activities would offer an additional strategy against the effects of labour-saving devices. It would definitely have a positive labour-market effect and at the same time offer an instrument for improved individual qualification.

However, we should bear in mind that reduced working time is an instrument that can be used only on a step-by-step basis, since unions need to insist on full pay compensation. Time reduction should be considered as an instrument that can be useful and helpful in improving the labour market situation. It is not an instrument that can be used so fast that all employment problems can be solved. Therefore, additional policies to prevent increasing unemployment are needed.

Qualitative Growth

The second suggestion would be to influence and to speed up the rate of economic growth. But there are different ways of doing this. One aims at quantitative growth. In this case the government spends money to stimulate the economy without caring for what happens in qualitative terms. Since industrial societies are prepared to accept many thousands of deaths per year through traffic accidents, we can afford to operate with a very cynical theoretical example. Let us assume that the government pays a premium to those who get involved in a car accident. This would certainly increase the number of accidents and consequently would increase the sales of new cars, the repair of damaged cars, the number of funerals, the sales of flowers, the jobs of doctors, hospital personnel, pharmacies, and many other sectors. National growth would go up. A society that cares only for quantitative effects is doing just this: not necessarily by stimulating additional traffic accidents but by stimulating energy consumption, waste of resources, increasing pollution, and so on.

The alternative would be what is called qualitative growth. In this case, the government spends public funds for the stimulation of activities that contribute to the improvement of the quality of life. But in this case it is important to direct industrial production into new areas. These areas must be future oriented and adequate to the need of the 1980s and 1990s. These needs are very well known. They cover products and systems for problems relating to catastrophes, pollution, accidents, education, public and private services, infrastructure, saving of energy and other resources and increasing the quality of working places and, most important, housing. All these problems can only be solved by using mostly industrial products. However, manufacturing industries are still oriented to the sixties, the period of purely quantitative growth.

The manufacturing industry needs to be completely restructured to meet the problems of the future. Only under these circumstances will it be possible to offer new additional jobs for persons replaced by microelectronics or other types of technical change. Microelectronics will contribute in many different ways to qualitative growth. It is a very important energy saver, for example. However, it cannot solve the energy problem alone. Therefore, additional activities are necessary. But there are strong reservations against this type of policy. These reservations are based on ideology, on financial restrictions and on the fact that the present instruments to run an economic structuring policy are still rather limited.

It does not make much sense to discuss ideological questions in this respect. However, it is true that instruments for structuring policies need to be improved. Therefore, scientific research in this area should be encouraged. Governments should make funds available and found new institutions to increase the theoretical and empirical knowledge of structuring policies.

The financing of a policy for qualitative growth is no real problem in most industrialised countries. Nearly all nations are spending extremely high subsidies for certain enterprises and branches of industry. Most of this money is used to stabilise and to conserve old structures instead of creating new ones. If the money available for subsidies would be used to a greater extent for future-oriented activities, the creation of new jobs through qualitative growth would be no problem.

In most industrialised countries, however, there is no co-ordination between research and development policies and other economic activities. The option of "qualitative growth" can be realised only if governments are able to co-ordinate all their very different types of policies (budgetary, economic, labour market, structural, R & D, housing, education, etc.) to attain one goal: full employment.

At least the industrialised countries should consider that for them unemployment is a very expensive matter. They should ask themselves if they are wealthy enough to carry this burden. They do not only have to pay unemployment compensations. Since unemployment is concentrating on relatively weak groups of the society, such as:

— young people with low education,
— handicapped persons,
— women,
— older people,
— all other groups with low education,

additional expenses for education programmes, for fighting against drug or alcohol consumption, are necessary. But this is not all. The unemployed do not pay taxes nor contributions to social security and they are very restricted consumers. So it would be more economic or cheaper to spend some extra money for employment programmes.

All investigations about the chances and risks of the 1980s show that the so-called "status quo policies" offer no solutions to the problems described. And, at least up to now, the British economic policy between 1979 and 1981 has not proved to be a useful alternative for human beings.

7

A Third World Perspective

JUAN F. RADA

Technology has played a crucial role in the economic, social and political developments of the past two centuries, both within national and international contexts.

Changes in technology deeply affect the worldwide distribution of productive facilities and services—the international division of labour. The Industrial Revolution is a classic example of technical and socioeconomic change which has profoundly altered human life. Early technological advances in textile manufacture, for example, changed the international division of labour. By the 1980s the price of yarn was perhaps one twentieth of what it had been fifty years earlier, and the cheapest Hindu labour could no longer compete in either quality or quantity with Lancashire's mules and throstles.

Similarly, the commercialisation of the Haber-Bosh nitrogen fixation process in the 1920s did away with one of Chile's most important sources of external revenues: the exploitation of natural nitrate. At one point, Chile supplied at least two thirds of the nitrate requirements of the world, and the tax on exports was 80 per cent of Chile's total revenue.

These examples illustrate the interaction between technical change and the international division of labour. However, as explained in earlier chapters, the characteristics and nature of microelectronics create a rather different situation today, one which challenges the traditional conceptualisation of technology, development and industrialisation.

Today, the importance of technology's role in development and the contribution it could make towards a more equitable world system are commonly recognised. The control of technology often means the control of development, the definition of its aims and even its pace. In these circumstances, the debate on technology should be placed in a wider framework, one which encompasses the very essence of commonly practised development strategies and styles.

Disparities Within and Among Nations

In order to assess the impact of current technological change, it is necessary to look at present disparities within and among nations. In the case of microelectronics or, more accurately, information technology, at least three different areas of disparity are apparent. The first is in the international distribution of scientific and technological capabilities. According to the United Nations Educational, Scientific and Cultural Organisation (UNESCO), the share of developing countries in total world expenditure on science and technology is around 3 per cent and they possess only 13 per cent of all the world's scientists and engineers. Even then, the lion's share of this expenditure and capability is concentrated in a few developing countries, such as India, Brazil, Argentina and Mexico.

The second area of disparity relates to industrial capability. In 1975 a conference of the United Nations Industrial Development Organisation (UNIDO) adopted an industrialisation goal for the Third World which became commonly known as the Lima Target.

In brief, it stated that the developing countries should increase their share of world industrial output from 7 per cent to at least 25 per cent by the year 2000. By 1980 this share hardly reached 9 per cent and, in the absence of special measures, will not exceed 13 per cent by the end of the century. In five years the Lima Target has been cut by almost half.

The third area of disparity is in the information infrastructure of society: the system that binds together different activities of a social, cultural, political and economic nature.

Many of the disparities in this field have been researched by the UNESCO programme on the International Information Order. In developing countries, 1 person out of 30 gets a daily newspaper, and only 1 out of 500 has a TV set, while in the developed countries the figures are 1 out of 3 and 1 out of 12, respectively. In addition, 83 per cent of the world's books are produced in the advanced countries. These disparities are now accelerating at a rapid pace as the information infrastructure becomes more dependent on electronics (see the chapters "The Technology" and "The Technology Applied").

In terms of the value of data processing equipment, the consulting firm Diebold (Europe) estimates that the United States, Japan and Western Europe account for 83 per cent of the world total in 1978. The 17 per cent share held by the rest of the world will have only risen marginally to 20 per cent by 1988. Most of this figure is accounted for by Eastern European countries and by some developing countries. Large Western banking firms possess more computer power than the whole of India. During the period 1978-1988 the "gap" in the value of equipment between the Western advanced countries and the others is expected to grow by a factor of more than two.

In terms of telecommunications equipment, the developing countries represented 10 per cent of the world's telecommunications equipment market (including telephones) in 1980 and will represent only 14 per cent in 1990. Already in 1980 the value of the world's electronic market was roughly equivalent to the combined Gross National Product (GNP) of Africa

In introducing a Third World perspective, it is also necessary to point out the great disparities within and between developing countries. The term "developing countries" is a shorthand concept describing many different levels of economic, social, political and cultural development, though often with similar colonial history and many common interests. These differences make generalisations hazardous and possibly misleading. Many of the issues related to advanced countries, which are treated in other chapters of the book, can also be applied to some of the developing countries, particularly the most advanced ones.

The array of electronic devices and products will probably only reach a tiny minority, the urban upper class and sectors of the middle class. The overwhelming majority of the population will not feel the impact nor reap any benefits. While the more advanced world is moving towards the integration of thousands of elements per "chip", in Africa only 1 person out of 18 has a radio; the transistor revolution (which spearheaded today's electronics nearly 30 years ago) has not yet arrived. All this points to the fact that the underlying socioeconomic, cultural and political reality of the world is not only increasing in complexity, but also widening the gap between the rich and the poor. Technological fixes of whatever nature are nothing but a drop of water in the sea of reality. Current trends must be redressed and this is becoming increasingly urgent as the advanced countries push inexorably towards a different societal texture. Although there is no easy diagnosis or answer to the dilemmas posed by rapid and different technological changes for developing countries, there are some fundamental issues that need urgent attention if a proper perspective is to be acquired.

Two other important elements must be mentioned in order to adequately assess the impact of technological change on developing countries. The first relates to the increasingly tough attitude Western countries are adopting towards the Third World. This results from the painful domestic consequences of their own process of adjustment. They have become less willing to take a global view of their own problems. This attitude has been manifest in the last two years and is likely to be accentuated, due partly to the impact of microelectronics on their own societies. The second element is that the expansion and development of microelectronics require a world market, a global strategy. This is a direct consequence of the research and development cost and capital investment needed to make companies truly competitive. Although the developing countries' market is in many cases very small, it is important for the overall feasibility of investment: and thus for the diffusion of the technology even in developed countries. Underlying this is the fact that electronic component, computer, and telecommunication technology is concentrated in a few transnational companies. As the productive structure of national economies becomes more dependent on these technologies, their power, already great, will increase, leading towards further transnationalisation of the economy and greater dependence. All the issues related to transnationals and transfer of technology will, therefore, become increasingly important.

The Impact of Microelectronics on Developing Countries

There are the following main areas where developing countries will be affected by microelectronics. These can be summarised as follows:

Comparative advantages

Comparative advantages are increasingly conditioned by science and technology, rather than geographical location or historical circumstances. This is not to say that the latter aspects will not remain of crucial importance, particularly in relation to raw materials and energy. But as far as production in the general sense, meaning not only manufacturing, is concerned, comparative advantages will increasingly be determined by scientific and technological know-how.

Traditional industries are moving towards a high technology category, heavily based on research and development and science. Science and technology are becoming the underlying base of industrial production, determining its structure and output, and therefore are increasingly the base of wealth creation. This shift will increase the gap between nations, given the present imbalance in science and technology—capabilities mentioned earlier.

Labour costs

In a number of industries the decreasing importance of direct labour costs in total manufacturing cost has reduced the developing countries' comparative advantage of abundant low-cost labour. This is the continuation of a historical trend which was first clearly evident in agricultural production.

The location of manufacturing facilities is not determined by labour cost alone, but by a number of factors—the type of product being manufactured, transport, flexibility, tax incentives, start-up costs, tariffs, government and fiscal policies, other aspects of international trade and labour legislation. Labour costs in many cases are now no longer of overriding consideration.

A massive return of industry to the advanced countries is unlikely. But relative stability of the current international division of labour with some roll-back in specific sectors can be expected. This has serious implications for the Third World's plans for industrial development.

The time horizon in which changes will occur will undoubtedly vary from sector to sector, but the trend is definitely towards a systematic erosion of the advantage of low-cost labour. With automation the cheapest labour is no labour.

Production

Production is increasingly the result of capital investment rather than the use of labour. In the field of microelectronics and other advanced technologies, this is certainly the case. A classic example is modern capital-intensive agriculture, where inequal access to capital and inequality in the availability of all the conditions necessary for its investment has led to wide disparities within and between countries.

In developing countries the modern technologically advanced sector is an enclave, geared in most cases to export markets. This increases the dislocation between the modern industrial sector, the artisan sector and agriculture. The question of linkage between advanced, less advanced and traditional technologies is crucial, for new products and processes alter sub-contracting arrangements and maintenance patterns, while the use of flexible manufacturing systems has the potential of replacing mass production of standard products by the continuous production of small batches of individualised ones. The incorporation of skills into equipment concentrates them in the hands of the manufacturer, producing a skill-saving effect in the production process, although an increase in the programming and design stages.

A polarisation of skills takes place, which appears not only at the plant level but also on the international scale, insofar as the suppliers of equipment are mostly in developed countries. This leads to the growing danger of technological discontinuity as the advanced countries push equipment to obsolescence and discontinue the production and supplies of spare parts and systems. In most developing countries this equipment could remain economically viable for a long time, but unable to operate because of lack of adequate supplies. Past examples are many and need to be assessed in the light of accelerating technological change.

The increase in capital intensity brings forward the old question of technological innovation and employment. The utilisation of computers and information technology in developing countries could produce both an immediate loss of jobs and a lowering of the country's job creation potential. This will be particularly true if the countries are not producers or assemblers of equipment and thus no compensation takes place within the national economy.

However, the impact on employment can be expected to be less important, in relative terms, in the Third World than in the developed countries. On present trends, other technologies such as extensive tractorisation and mechanisation of agriculture are of a much greater importance for employment. More than the direct utilisation of the technology, the erosion of comparative advantages and the change in the nature of the productive infrastructure have a far greater effect on employment, especially when seen in the context of the calculation by the International Labour Organisation, which suggests that 625 million jobs need to be created in the Third World between 1975 and 2000.

Integration of functions

The increasing level of integration of electronic components transfers functions previously performed by mechanical, electromechanical, electric or older electronic components into single components. This concentration of various functions into one single component has important consequences for developing countries. A primary strategy for the absorption and development of technology by developing countries has been the "unpacking" of technology: the division of its various processes and products into separate parts and sub-stages. This now becomes increasingly difficult as technology is "packaged" or "frozen" into the components themselves, becoming a "black box" for the user. The "packaging" of technology into single components produces a chain of effects. Firstly, the value which was previously added in the process of manufacturing and assembly is transferred to the manufacturing of components. Secondly, the grouping of a growing number of functions into single components compels system design.

This transformation further reduces the capacity of developing countries to absorb technology, particularly at the application stage. The alteration of managerial requirements also reinforces one of the traditional advantages of developed countries. Developing countries are further affected by the obsolescence of many mechanical skills.

The importance of information

The production of goods and services is increasingly dependent on the efficient processing of and access to information. The so-called "information intensive sectors", i.e. data banks and other data bases, data processing and information services of a scientific, technical, commercial or historical nature, are heavily concentrated in the developed countries. The questions that these rather new phenomena pose are of great importance for developing countries economically as well as culturally. For instance, the use of data bases implies the need to import services and in most cases the hardware to receive the service. To the extent the values embodied in information are not independent of its format and system of storage, this will also involve the import of cultural patterns and interpretations.

Infrastructure and human resources

These five main areas of concern are less applicable to those countries which do not possess an industrial base or even an incipient one. But they should be aware of policy options and possess an adequate understanding of the dynamics of industrialisation. A precondition for the utilisation of information technology is an adequate infrastructure, regular electricity supply, and a fairly sophisticated telecommunications infrastructure (which in many advanced countries is taken for granted). Useful and practical applications of microelectronic technology in the least-developed countries can be found particularly in agriculture. Nevertheless, under the conditions prevailing in this large group of countries, the gauge by which to measure the positive impact of the new technology is the national capacity to absorb its applications, rather than the simple importing of complete systems. Furthermore, if advanced technological innovations increase the inequalities within countries, they will hinder rather than help development, as happened to a great extent with the "green revolution".

Technological Changes and the Opportunities

Although the problem of adjusting to their negative effects presents considerable difficulties, current technological changes also offer many opportunities. The incorporation of skills into equipment, such as in precision machinery, will save skills which are often difficult and costly to acquire. In the case of some precision engineering activities where labour is required in great quantities, new opportunities to leapfrog traditionally low- and medium-grade mechanical skills have become available.

At the same time, the full utilisation of human resources could bring important gains. Low-cost, highly educated labour can be used in areas such as the conversion and development of software which is likely to remain labour intensive. In this respect economic and educational opportunities are open, particularly if software is utilised in the wider context of science and technology policy.

Information technology, particularly interactive links, can be used to increase the flow of scientific and technical data within and across countries without changes in geographical location and therefore help to create the "critical mass" necessary for innovation.

Both aspects depend on public policy and are important for absorbing current changes, maintaining cultural identity and obtaining some degree of independence in the scientific and technological field. The cultural importance of information technology needs no emphasis; unless developing countries pursue a forceful policy and direct the applications of the new technology for their own purposes and aims, their level of cultural dependence will increase.

Many other aspects of current technological innovation could be used to great advantage (e.g. remote sensing for agricultural development or direct broadcasting for education and other purposes), but again it will depend on prospective policy, the full utilisation of local human resources and pooling of resources on a regional basis.

The realities of current and future technological change show that national approaches to application, production, even assembling of equipment, and development of know-how are in most cases clearly insufficient. A broader regional approach is required.

Nationally, there is no sufficient body of research and development experience, economy of scale, financial strength or market to make advanced technology viable. Some of the larger developing countries could standardise components at a lower technical level, since this will meet most of their needs. In this case, a conscious policy of decoupling from the international market needs to be followed in order to develop an industry behind protective barriers. Even then, such a strategy is only feasible where the internal market is sufficiently large.

On the following pages specific aspects of certain sectors will be briefly analysed in order to illustrate the assertions made earlier.

The Electronic Industry

The electronic industry is a very heterogeneous one, covering many different types of activities. This section will mainly deal with the component sector to examine how the shift towards more complex production processes erodes the advantages of developing countries.

There is a clear understanding in developed countries that microelectronics is essential for the maintenance of their industrial base. This has been summarised by the Advisory Council for Applied Research and Development in the United Kingdom, which argued that unless the country embraces semiconductor technology, it will soon "join the ranks of the underdeveloped countries".

For developing countries, there are two areas of concern in relation to current developments in the electronic industry. Firstly, the immediate effect (particularly on export-oriented economies), and secondly, the medium-term prospects of policy options and capacity to absorb the technology. Both these concerns must be seen in the context of the speed at which innovation is taking place and the changes that this has produced at the product and process level.

It is evident that different countries follow quite different development strategies and any generalisations are hazardous. This section will concentrate on a handful of developing countries which have followed a strategy of export-led growth. This is not typical of developing countries, but it has been chosen because it illustrates in a very forceful and clear manner the industrialisation dilemma facing developing countries. To a greater or lesser extent, all countries wish to participate in the world market of semi-manufactured and manufactured products, thereby diversifying away from agriculture and raw material production. The route followed differs from strict import substitution with protective barriers to fairly open economies which use the mechanisms of free zones for offshore installations of foreign companies. Many developing countries have adjusted their legislation precisely to allow this system to operate, particularly in South East Asia.

For these countries, electronics has been a substantial element in their overall strategy since, in the past, the characteristics of the industry called for extensive use of labour in the assembly phases and required low amounts of capital per worker employed. These countries have upgraded electronics in an effort to diversify towards high quality products, increasing the level of nationally manufactured components in the finished products.

Transfer of technology

The highly competitive environment in which the electronic industry has evolved forces a policy of constant cost reduction, rapid innovation leading to shorter product cycles and a powerful marketing infrastructure. The cost reduction policy led to the extensive use of offshore installations during the 1960s and early seventies for the assembly of components and other items. This was partly related to the U.S. special tariff arrangements.

In the past all major manufacturers of electronic components used offshore installations for the wiring and encapsulation of "chips". Since about 1978, however, the trend has been to open the new generation of plants in the advanced countries. There are a number of reasons for this: automation, the need to be closer to end users, quality control, elimination of logistic problems, and the need to upgrade local or national capabilities.

Since 1978, the major investment in semiconductor plants has been in Europe, Japan and the United States. At the same time, offshore installations are being moved from high-cost Asian countries, such as Singapore, Hong Kong, Taiwan and South Korea, to countries like Thailand and the Philippines in an effort to cut cost further. In other cases automation of offshore plants is taking place.

A number of factors combine to reinforce this trend ranging from the policy of industrialised countries' governments to ensure a microelectronic base, to protectionist measures and currency values which affect, for instance, Japanese producers in their effort to export to Europe. Economic factors and the political climate affect together investment decisions. But it is important to realise that in the case of electronic components, the growing integration and the automation of the manufacturing process is a crucial element. Since labour costs are not an overriding consideration, this process allows greater flexibility in investment decisions.

The strategies of each manufacturer are different and due to the characteristics of the industry are constantly being modified. For Japanese producers, the use of offshore plants represent only about 10 per cent of production and in some areas (discrete semiconductors) it is declining. Japanese automated bonding equipment represents a labour efficiency factor (in terms of people) of ten to one. This has helped to compensate for growing salaries in Japan. Another factor of equal importance to Japanese producers is the concern for quality which they feel cannot be maintained through the use of offshore installations. This element is growing in importance and will force other producers to follow similar considerations.

Countries which have relied heavily on foreign investment and imported technology in the electronic industry are becoming more vulnerable. They have been criticised on a number of occasions precisely because of the dependence of their economy on outside decisions, in particular in relation to the type of technology, its domestic diffusion and the product-cycle of foreign investment.

The rapid pace of change could imply technological discontinuity with severe effects on the industry. Products which remain labour intensive in the assembly stage have lost an important part of their value-added elements due to the change in components. This is the case, for instance, with TV sets, where the number of parts has been halved since 1970 because of the use of more sophisticated components which have eliminated many assembly operations.

Today, advanced electronic technology—mainly concentrated in U.S. and Japanese transnational firms—is obtainable for the most part only through coproduction, joint ventures and cross-licensing; however, generally more than one channel is used.

Transnationals tend to make no systematic effort to encourage local research and development, or to reinvest profits beyond the minimum needed to maintain the plant running. Governments therefore need to negotiate with companies for greater local R & D content, while at the same time encouraging basic research and a closer link between technical institutions and local firms. As technology becomes the basic asset of companies, their willingness to transfer it will decrease even further and turnkey projects will increase.

Transfer of technology is possible only if the absorption capacity exists from the human resource side. Even developing countries which are advanced in electronics, such as South Korea, had only 0.4 researchers per 1000 of the population in 1978 compared with 2.6 and 2.4 in 1977 for the United States and Japan, respectively. Expenditure in R & D per researcher is about $21,500 in Korea, compared to $47,560 in Japan and $80,680 in the United States. Figures on R & D do not tell the whole story, since the process of innovation is a complex one where multiple factors condition the final results. Furthermore, the cost required to shape an innovation into a marketable product can easily be as high as 1 to 20; on top of which a powerful marketing structure is also required.

There is no doubt that as time passes, some developing countries will be able to absorb part of the technology in specialised areas. Whether that will give them any significant stake in the international market for components and equipment is another matter.

For developing countries which have followed a domestic oriented pattern, the alternatives, in the light of current changes, are different. The trend has been to diversify dependence in the component field and assembly of systems and peripherals. Some choose to standardise at a lower level by manufacturing wafers for the domestic market rather than try to compete in the international market.

R & D and capital investments

As the speed of technological change increases, so do the requirements for capital and research funds. According to figures from the Organisation for Economic Co-operation and Development (OECD), already by 1975 industries producing electrical and electronic machinery, equipment and supplies (including computers and data processing supplies and accessories) were the largest users of R & D, outstripping the chemical and aerospace industries.

Many other areas not accounted for in the electrical groups, such as aerospace and machinery, are in fact heavily based on electronics and considerable R & D efforts are undertaken. In addition, production support, in-company training, management development, and other items which are part of the "technological environment" are not accounted for, although they may have a significant impact on results.

It was around 1975 that rapid technological change and keen competition led to an acceleration in the use of R & D by the electronics industry. The ten most important U.S. "independent" semiconductor companies (the companies that sell on the open market) more than doubled their R & D budgets, from over $200 million in 1975 to over $550 million in 1980. These figures include expenditure by the companies and exclude programmes from outside contractors such as the U.S. Government.

An element which is not considered here, essentially because of lack of data, is investment in basic scientific research. This is becoming increasingly important to maintain the lead in electronics.

The next crucial element in generating advantages in electronics is capital investment. Electronics can no longer be seen as a labour-intensive, low-capital investment, except in some assembly operations which are becoming less important as the industry reaches for global markets and economies of scale.

An indication of the changes in production costs is the fact that a typical production line of integrated circuits in 1965 cost about $1 million. In 1980, due to the complexity of the circuits and automation, the cost ran as high as $40–50 million. The combined effects of growing R & D and capital expenses have resulted in business, and thus technological, concentration which affects developing countries by a further polarisation of the international division of labour.

Some plants will stay in developing countries. Their task will be mainly the encapsulating, wiring and testing of components, but sophisticated wafer fabrication will remain in the developed countries, particularly in the United States and Japan.

There are three main reasons why a massive repatriation of offshore assembly in this sector is unlikely in the medium term, although the amount of investment is and will diminish further. First, under financial stress companies prefer to maintain offshore installations and invest at a slower pace in new plants, thereby retaining reduced profits from the installations without incurring the cost of considerable and unjustified new investment. Supply problems can also extend the life expectancy of offshore installations. Increased salaries in South East Asia, however, are forcing some companies to search for different locations, while those staying in countries such as Singapore are increasing automation of assembly.

Secondly, labour legislation in offshore locations often permits working conditions that are not tolerated in the developed countries.

Thirdly, there is great pressure by governments in South East Asia to maintain facilities and upgrade technology. Direct government intervention and guarantees are a very important element. A number of agreements already reached in some countries for wafer fabrication, aimed at the lower level of the semiconductor market, are the result of such incentives.

In the case of consumer products (TV, for example), the continued use of offshore installations will depend very much on what type of product is manufactured. Thus, for those products where the selling price and margin are low, the assembly time long and the relative weight of labour in the product high, offshore installations will be maintained for some time. However, in those cases where the product incorporates expensive raw materials, is produced in long production runs, and has a short assembly time, it may be profitable to repatriate the installations from the Far East. With automation and robotisation, the cost of investment and utilisation remains basically the same, regardless of location.

Software and man–machine interface are areas where some developing countries could play an important role, providing they possess a degree of specialisation and forceful govenment policy. It must be understood, however, that an important level of dependence will continue to exist at the component level. This should encourage developing countries to pool resources in order to reach economies of scale on a regional basis, even at the software level. Software is a complex skill, but at the same time remains an extremely labour-intensive activity which demands close contact with the end user for most applications. India with its large portion of highly educated people and its low labour costs, has successfully been able to export software to developed countries.

The Garment Industry

An interesting example of how technological change is conditioning the ability of developing countries to compete is garment manufacturing. Some developing countries have based a major part of their industrial development strategy on textiles and clothing because of their high demand for labour, local raw materials, traditional techniques and low investment/output ratio. In addition, they permit import substitution and eventually exports. Clothing represents about half of total manufacturing exports from developing countries to the developed ones and this explains why changes in this area are important in terms of the international division of labour.

The garment industry, traditionally, has been very segmented and one of the least automated and most labour-intensive industrial sectors. Garment manufacturing is typically composed of a recurrence of different activities with connected manual operations. It is estimated that about 80 per cent of the direct operative labour force spends 90 per cent of its time picking up, positioning, manipulating and removing pieces around stitch-making and cutting operations.

Most of the progress in the last twenty years has been on the stitching cycle, rather than on handling. Changes in stitching have been enhanced by microelectronics because of more reliable and powerful industrial sewing machines, where hundreds of components have been replaced by "chips". In addition, microprocessor-controlled sewing machines are able to monitor the temperature of the needle and regulate the speed accordingly, in order to avoid damaging the cloth or causing a breakdown.

In the initial cutting and marking process which has been traditionally highly skilled and labour intensive, computer technology is today widely used by the larger manufacturers. Since as much as 50 per cent of the cost of garments is the cost of the cloth, optimising the utilisation of the material is crucial. Optimal pattern placement is worked out by the computer, which also guides a laser beam cutter. With this equipment, the reduction in labour can be dramatic, in some cases reducing the number of people required from 200 to 15, with savings on cloth wastage ranging from 8 to 15 per cent. A number of edge guidance devices are now available. These automatically monitor the edge of the material to be sewn and draw the sewing foot along at the same time.

These are part of a growing number of applications, leading towards a "total system concept". This means the use of computerised equipment to detect flaws, optimise layout, keep track of patterns and orders, monitor the progress of work throughout the plant, automate the matching of patterns, the cutting and sewing.

Important technical changes are also taking place in the assembly stage, where the separate components and subcomponents (collars, buttons, pockets, ornamentation, etc.) are put together. This is one of the areas where subcontracting to low-wage companies in developing countries often takes place. Later on, automation in the handling operation will greatly diminish the labour content at this stage.

In the high-wage developed countries the weighted average annual cost per worker is about $12,000. In the main developing countries the average cost is about $2000. This difference is so great that it is doubtful whether new technology can offset it if the technology is simply used to reduce the importance of labour costs. This is certainly a crucial point. However, the new technology permits better and more flexible design, fabric and quality improvement, customer service, product innovation and risk reduction, which is important for the marketing and "appeal" characteristics of the garment industry. Thus, a wider understanding is required in order to increase the efficiency not only of labour, but also of the other elements mentioned above.

The wage differential and the segmentation of the industry suggest that the pace of change will be slow and different in each segment. It also implies that many areas will remain in the developing countries. The technological trends imply a decline in the labour content of the garment industry. At the same time a more skilful and flexible labour force will be needed with growing skill requirements at the software level and also an increase of managerial and office productivity will be required.

The latter aspect is of great importance, since an increase in "white collar" productivity could be as significant as changes on the shop floor. As the industry becomes more and more dependent on managerial capacity and complex information flows among the different stages, it will further reinforce the traditional advantages of developed countries. Furthermore, the extensive use of for example computer aided design will link in with the manufacturing process and will increase automation.

The technical and economic problems of automating and mechanising garment production remain formidable but not insurmountable. The pace of innovation has already been accelerated by the combined effect of the existence of reliable control elements and other technologies as well as the economic and social pressures generated by the need to perform adequately on the international market.

A report for the European Economic Community (EEC) on apparel manufacturing argues that if available technology is developed to practical application, it could make a significant impact on competitiveness to the advantage of the manufacturers in the high-cost countries. The report concludes: "The size of the advantage enjoyed by the low-cost countries is such that technology alone can in no way be regarded as a panacea, but it can and must be used to erode the significance of that advantage if the industry is to survive."

If the garment industry pursues the technology path, there will be a considerable change in the structure of the industry and its capital requirements. In the short run, automation will be concentrated in high-quality fashion products, where developed countries have a lead-time of six or more months. Current protectionism in the industry, based on import quotas, has forced some producers in developing countries to move up-market in order to increase value while reducing volume; this also increases their vulnerability. Furthermore, the labour cost differentials are eroding as some countries industrialise.

Although it is too early to evaluate the final consequences, the trends seem to point to large-scale concentration in the garment business in the medium term. This trend is already discernible in Western Europe, the United States and Japan as companies search for the economies of scale necessary to justify large capital investments.

The garment industry is living in the dawn of a transition towards different types of skills, higher investment and concentration. The cost of implementing the technological changes required is pulling together the manufacturers of equipment and end users, making the industry at the same time more dependent on research and development. It is worth noting in this context that breakthroughs in automatic cutting came from aerospace and associated large-scale high-technology enterprises.

The technological and business trends in garments suggest that the diffusion of the technology will be uneven. The very basis of the industry is geared towards the "high-technology category" and hence to a reliance on software and systems skills, complex maintenance skill (in such areas as optoelectronics and robotics), design and managerial capabilities.

The combination of these factors could give the producers in the developed countries an advantage which will allow them to recuperate important segments of the market and retain part of the industry in the North, albeit with reduced manning levels.

In the medium to long-term—with a time horizon of 5 to 10 years or less—the competitive edge in garments will cease to be labour and will increasingly be technology in the wider sense as defined earlier.

Information Flows and Developing Countries

The nature of microelectronics in its capacity to process information affects information flows within and between countries. This has been shown in the chapters "The Technology" and "The Technology Applied" and is explained in further detail in the chapter "Information Technology and Society". The economic aspects and consequences for the Third World of such a process need to be emphasised. Much of the computer power and information processing industry is in the advanced countries, particularly in the United States. This concentration of "information intensive sectors" or informatic power in few countries reinforces the imbalance between developed and developing countries, which is evident in the trade of tangible goods.

It is not possible at this stage to determine the exact value of the flow of data across borders, since it involves many activities of different kinds. Evidence suggests, however, that it has significant importance and will increase considerably in the next ten years.

There is also the cost of importing computing services, which has not only an economic consequence but also involves a loss of control over the direction of future economic and social developments by the exodus of key decision-making processes. For most developing countries, the pressure on the balance of payments relates to the import of a service and the hardware and infrastructure required to receive the service. Furthermore, due to the decreasing cost of communications and the concentration of "information intensive" sectors in developed countries, it is becoming cheaper in many cases for Third World enterprises and institutions to send their design problems, calculations, research and routine data abroad rather than assemble and develop local teams.

Developing countries tend to use the data processing facilities of the industrialised countries for several reasons: routine data processing can be done more economically, the data centres may possess expertise not available locally, and their data bases contain vital information not obtainable in the Third World.

What happens, in fact, is a sort of "electronic brain drain" resulting in the reduction of opportunities for local development. The ever-growing dependence of many countries on large data networks will increase the importance of these issues. The economies of scale achieved by information centres in developed countries makes them very competitive internationally. Brazilian officials have stated that "dumping" of cheap time-sharing services has discouraged the establishment of this type of service in Brazil.

Transborder data flows have created the potential for greater dependence and with it the danger of losing legitimate access to vital information affecting economic and social developments. This has important political implications. The risk of retention or selective release of data is real. Under conditions of economic war, deterioration of the international political climate or indeed by unilateral decisions, a country can withhold data with harmful consequences to others. Some balance between imports and exports of data is required and this fact needs to be considered by countries which are just establishing international data links.

The concentration of data processing facilities, data banks and data bases poses the question of the distribution of and access to information. Although one of the advantages of current changes is the access to vast pools of information, disparities in this area will continue to grow.

In the scientific and technical field greater diversity is required in order to accommodate the needs of developing countries in areas as diverse as traditional medicine, rural technology and education. In this respect there is also a tremendous need for cultural diversity, not only linguistically but also in terms of content and format. Developing countries should collectively implement the storage of the "collective memory" in their own language and according to their own interests.

Diffusion of information technology

The diffusion of information technology in developing countries resembles the general pattern found in Western countries, but has its own characteristics. The most visible aspect of this diffusion is the growing use of computers, although other types of equipment are also introduced.

Today the computer population and telecommunications capacity in developing countries are only a small fraction of the world total. At the same time, many products, notably capital goods, are being transformed by microelectronics and this in turn is influencing manufacturing processes.

There are several kinds of practical factors that condition and slow down the rate of diffusion. One important factor is labour costs, particularly in direct application of computers for conventional procedures. The low cost of labour makes the equipment less competitive and amortisation takes longer. The lack of standardisation in a number of activities and of modern managerial practices has made software requirements more demanding and, therefore, the total cost of installations more expensive than they would have been otherwise. "Software packages", even in the banking sector, are difficult to implement. This situation has led, paradoxically, to well-developed software applications in some countries. Additionally, companies need to spread the costs of the infrastructure, services and operations over relatively few pieces of equipment, which also contributes to increased costs.

Most computer applications in developing countries are of a traditional nature, that is for payroll, accounting, stock, routine administrative work, billing, etc. Although this is the same pattern as followed by the West, inasmuch as an essential core of specialists and skills is developed, some developing countries can expect a more decisive move to less conventional applications in the future. As in the West, some sectors such as banking, insurance, finance, large companies, utilities, airlines and governments, use computers more intensively.

The role of trade union movements has not been significant in acting as a brake, although examples of industrial action over computerisation exist in several countries.

The single most important factor concerning the diffusion of the technology is undoubtedly government action. This has occurred in a wide range of fields, from imports, telecommunications and transfer of technology policy to industrial strategy. An increasing number of countries have adopted centralised systems to process import requests, monitor applications and purchase equipment for government needs. These types of measures cover especially data processing equipment and do not affect other types of electronised devices. The situation in each country varies considerably.

India, for instance, manufactures equipment under licence (about 30 per cent imported components) and has introduced regulatory measures on foreign investment in the field. Given the current and potential impact of information technology, it is likely that an increasing number of developing countries will combine regulation on imports and applications, following a path similar to India. At the same time, the penetration of the technology is partially due to the marketing drive of the manufacturers, who are under constant pressure to open new markets, mainly because of the rapid technical obsolescence of hardware. This compels them to sell or rent as many computers of a given generation as possible. The technical obsolescence criteria of the developed countries is almost mechanically transmitted to the developing countries, leading more often than not to an unnecessary frequency in the upgrading of equipment.

When all factors are taken into account, it is evident that a relatively rapid diffusion can be foreseen in computers and telecommunications and in some industrial applications. This, however, does not necessarily imply that developing countries will be able to absorb microelectronic technology at a pace that will put them in a position to compete with the industrialised countries. They lack the awareness, the skills and capital requirements. In this respect developing countries are bound to benefit less from current changes.

Employment Prospects

Although the effect of information technology on employment has been in the centre of debates in the developed countries, the issue is rather different for developing countries. This is due to the structure of the labour market and to widespread underemployment. More important than the direct utilisation of equipment, which affects a small proportion of the labour force, is the narrowing of industrialisation alternatives resulting from the erosion of comparative advantages. Many factors other than technology are important in determining these changes. But technology does increase the options for developed countries. It reduces the importance of direct labour costs as a result of changes in products and processes, while at the same time increasing flexibility, quality and overall productivity and performance.

In some industries developing countries will be chosen as sites for establishing new plants based on the new technology. The decision will be based less on low labour cost than on tax advantages, lack of labour resistance, lower start-up costs and cheaper capital. It is doubtful, however, whether this type of modern installation will create significant employment or contribute to the general development of the economy. The diffusion of microelectronics in developing countries implies a sophisticated technology for most products and processes, employing few people.

At the same time, the use of the technology by developing countries will widen the gap between technology and its application, because entire processes will have to be imported on a turnkey basis. The net result could be islands of high technology within economies characterised by low-productivity and artisan-based production systems. This dichotomy, and the interlink between the two areas, represents one of the existing dilemmas of industrialisation.

The qualitative characteristic of industrialisation are as important as the measurement of industrial output, for they have consequences for income distribution, employment and the general structure of the economy.

The industrial and service structure in most developing countries is sharply divided between a modern sector and a multitude of small businesses with extremely low productivity and high underemployment. From the point of view of employment, this "formal-informal" sector relation poses a qualitatively different problem. A reduction or loss of potential employment in the "formal" sector increases pressure in the "informal" one, provoking further underemployment or unemployment there. This reinforces the pattern of low productivity and further inhibits economic growth. This classic circular situation shows that the introduction of advanced technology has a chain effect and cannot be analysed only at the level of the individual enterprises using it. This aspect is particularly important when the equipment is imported and thus does not enhance internal manufacturing capabilities.

Studies undertaken in developing countries on the employment effects of computerisation tend to agree that there is a loss of potential employment, although there is some job creation in the short run due to the setting up of data processing departments. A similar pattern occurred in the early stages of computerisation in the advanced countries, but changes in input techniques (which are replacing key-punch operators) will diminish the level of job creation in the future. Furthermore, most developing countries import the equipment and thus no employment is created at the assembly or manufacturing level.

For some developing countries the combination of high technology and low labour cost, particularly of educated labour, could be one of the opportunities to develop an embryonic informatic industry and thereby to create employment. Already low-cost, highly qualified manpower is being used in a number of areas such as civil engineering, the provision of technical services in health, finance, agriculture and consultancies in general.

Whether these opportunities can be increased will depend on developments in the area of software. Software will remain for some time a labour- and skill-intensive activity, thus the low cost of labour can continue to be a comparative advantage in this area. Such an advantage depends on the country's current endowment of software skills and its capacity to keep software specialists at home. The development of capabilities in this area will largely depend on public policy, since in many cases the purchasing of skills outside the country is more economical than developing national skills.

The question of "brain drain" should not be overlooked, since the recruitment policy of large hardware and software companies is as global as all their other operations. Developed countries are already subcontracting software development to developing countries, especially for the software routines and applications which are most relevant to their conditions.

Information Technology and Development

Information technology is a reality, and a rapidly expanding one. Therefore, the question is how to master the changes and deal with the issues it raises to the best advantage for development strategies.

A socioeconomic command of the development of science and technology is needed. It is obvious that the issues being confronted today, which will increase in complexity in the future, demand a global approach. This is also true for science and technology. The implementation of such an approach will undoubtedly take many years, but it is unavoidable. With the increasing economy of devices such as robots, the possibility of complex direct verbal input, direct satellite links and many other changes under way, time cannot be wasted.

The essential starting point is the development of a system of scientific and technological assessment, scanning and policy design. The capacity for prospective assessment permits not only a better bargaining position but also puts the elaboration of development

strategies on a more solid basis by developing a more refined conceptualisation of short-, medium- and long-term comparative advantages. It will be necessary to learn how to harness current changes,
while avoiding the undesirable effects of technology. Microelectronics-
based innovations can be of great benefit, if properly applied.

Three general principles apply to the use of the technology within
countries. First is the need for a national policy based on a careful
selectiveness in applications. These should be aimed at overcoming
bottlenecks and optimising the use of resources, rather than using the
technology to replace labour, or to increase efficiency which could be
increased by other means.

Selectivity of application and conscious planning can, for instance,
minimise hard currency expenditure if data banks and data bases are
developed for domestic use and export. By avoiding imports of services, the foreign exchange savings could be greater in the medium
term than the cost of installing the equipment, while at the same time
local expertise and capabilities are also developed. It will also have the
added advantage of organising a system that meets the needs and the
culture of the country.

Additionally, selectivity implies a policy on technical obsolescence
and upgrading of equipment, together with clear technical criteria on
the types of equipment that could be used. This, of course, applies
differently to each sector and area.

Second is the need to assure diversified sources of supply in the
market and to avoid becoming dependent on a few companies. Such
dependence can greatly distort prices and application criteria because
of the companies' disproportionate marketing power. This must be
accompanied by a policy on software.

On the service side, the "technological package" should be
unwrapped in such a way that conversion software and other services
could be produced locally, providing an adequate manpower policy is
followed to train and develop local personnel and firms.

Diversification of the sources of supply is the only way to minimise
dependence, given the fact that it is highly unlikely that developing
countries can develop their own manufacturing capabilities. The
possibility of manufacturing exists only when the adopted criteria of
obsolescence differ from those of the international market. Even then

the system would have to be developed behind strong protective barriers. The technical specifications of the equipment would, however, be several years and generations behind the leading producers.

The third element is to monitor the national integration of locally assembled or partly manufactured electronic-based products. In general, countries use an index of national integration combining weight-volume-value to measure how much of a given product is nationally made in compliance with local legislation or plans.

Today integrated circuits, which are the heart of electronic-based products, have a very low weight-volume-value, but they incorporate all the "real" value of the product from the point of view of technology and know-how. If the purpose of a country is to upgrade slowly its own national capability in this area, it is important to monitor national integration in the light of changes in the technology under way. This is also related to the fact that many "intelligent" products appear under traditional or very general descriptions of import classifications or the Standard International Trade Classification (SITC) which may not only decrease the capacity of local industry to compete, but also implies that many products escape general guidelines and automation policy.

Possible economies

The capital-saving characteristics of the technology and its skill-saving effects allow "leapfrogging" of certain technological stages in some areas such as precision engineering. At the same time, traditional equipment is rapidly pushed to obsolescence, not because it is intrinsically outdated, but because developed countries need to use the latest equipment available to be able to compete among themselves. This equipment can become available at scrap value and remain economically viable when combined with lower labour costs, appropriate skills and appropriate managerial and government policies.

In many of these new areas of development, some skills are not difficult to obtain (conversion software, simple programming, etc.), although mastering the programming needs of standard types of equipment is not equal to the creation of innovative capacity and systems-engineering skills. The ''core'' of the software is generally part of the producer's package, and as the competitive edge will be increasingly on the software rather than on the hardware content, the shift to ''incorporated technology'' will grow even more in importance while becoming increasingly intangible. ''Reverse engineering'' in this context will be even more difficult. Transfer of technology, patents and licensing will therefore be much more complex, particularly since the product being sold is as intangible as nonmaterialised knowledge.

Social and government applications

Computer technology has long been used by government services in developing countries, especially for statistical purposes. This can increase further, improving the accuracy, reliability and timing of statistical information. This represents a valuable input into decision making and planning. The time lag between events or policy implementation and their proper evaluation can be shortened considerably, thus increasing the efficiency of decisions, policy design and the monitoring of the performance of different measures. With a solid information base, the technology can be used to optimise the allocation and use of resources, which in itself could mean considerable economies. A tighter control of commercial stocks, imports and exports, and tax collection could be of great benefit, while saving foreign exchange in many areas.

Although employment could increase marginally where new or supplementary services are created, in general the technology will diminish the job creation potential. Thus a careful evaluation is needed to combine traditional and modern methods with desirable information and employment results. Countries can further their planning and bargaining capacity if they are capable of assembling the information relevant to their interests.

Other important areas of applications are those which enhance social services, particularly health and education. In education, information technology can improve the capability of the traditional system, integrating remote and isolated sectors to national life and increasing the diversity and national content of programmes. For instance, the economy of audiovisual production and equipment can boost national programmes for schools, television, village education and rural extension just as the mimeograph boosted the local press and decentralised the production of educational materials.

A Call for Action

There are many areas where developing countries are called upon to act collectively. These could be facilitated by the use of information technology. These areas range from the co-ordination of industrial, agricultural and financial policy, which could allow massive horizontal links in the field of trade, to the strengthening of technological capacity and the development of adequate information and informatic systems. Collective action should be taken in three main areas:

1. Joint efforts to develop technological and scientific assessment and forecasting in those areas which are most likely to profoundly affect developing countries and the international division of labour. Social command of scientific and technological development is thus broader than mere control; it assumes an orientation of scientific and technological development rather than a regulatory and reactive position towards change. Existing institutions should be used for this purpose.

2. Joint efforts to evolve a common information policy. This policy should encompass data banks, data bases and networks in the economic, scientific and technological, research and development (R & D), cultural and mass media fields. The Third World should also develop a common policy towards communications, transborder data flows, satellite links, transfer of technology, and establish common facilities on a regional basis for manufacturing in selected areas, applications and R & D, with special emphasis, at least initially, on software.

3. Joint efforts to obtain preferential treatment for access to data banks and data bases as a way to mitigate the growing gap between developed and developing countries in areas such as science and technology. Free access to these sources could be part of agreements dealing with aid and transfer of technology.

It is necessary to warn that at present the benefits of the new technology remain largely on paper. To secure these benefits requires both short- and long-term policies. It necessitates an active search for alternative development strategies and this, in the final analysis, is related to the power structure within and between countries. If this structure is not altered in most countries and internationally, there is little hope that desirable benefits will materialise.

Data, information and a new productive infrastructure should not benefit only the few; we cannot have a world divided between information "poor" and "rich". Data and information should not be used to infringe on a people's cultural identity and invade, by means of different lifestyles, patterns of consumption and values, a world that is struggling to reach its own identity and development path. More than legislation and protocols, a new atmosphere of social command of technologies should be developed. In this atmosphere a participatory and pluralistic discussion about the use of technologies could take place. The wonders of current change could be used to solve pressing needs and for the benefit of all in a more interdependent rather than dependent world.

8

Microelectronics in War

FRANK BARNABY

In the past fifteen years or so, the characteristics of major weapons—tanks, combat aircraft, missiles, and warships—have changed beyond all recognition. This change is largely due to developments in microelectronics. Military microelectronic systems have revolutionised the guidance and control of weapons, and military communications, command and intelligence.

It would take many volumes just to describe all the military uses of microelectronics. Consequently, we will discuss here only four applications—in strategic missile accuracy and strategic anti-submarine warfare, in airborne warning and control system aircraft, in the automated battlefield, and in modern combat aircraft.

These applications are especially chosen to indicate the range and scope of the military applications of microelectronics, and to show the wide ramifications such applications can have for military tactics and strategy, for the future battlefield, and for the development of nuclear policies.

Military technological revolutions which have been, or will be, brought on by microelectronics are most usefully seen in the context of advances in military technology in general. These advances are made possible by military research and development. Military research and development, in other words, impels military technology and is, therefore, the activity most responsible for the arms race.

Military Research and Development

If there were no military research and development, no new major weapons would be produced and no significant improvements would

243

be made in the performance of existing weapons. The arms race, at least in the qualitative sense, would soon grind to a halt, even though the size of arsenals may increase and the arsenals of the smaller powers may be brought technologically closer to those of the great powers by the global arms trade.

A huge amount of money is spent on military research and development: today, a large fraction of it goes into the development of electronics for weapon systems. During the 1960s, an average of about $16,000 million a year was spent on military research and development, about 10 per cent of world military expenditure. Now, roughly $50,000 million a year is spent on military research and development, also about 10 per cent of world military expenditure. Even these huge sums are conservative estimates.

In constant prices, to take inflation into account, world military research and development spending increased by about 60 per cent during the 1960s and by about 20 per cent during the 1970s. Judging by the increases in military spending planned by some major countries, military research and development spending is likely to increase more rapidly during the 1980s than during the 1970s.

The money spent by governments on military research and development is a large fraction of the money they spend on research and development in general. In the United States, and almost certainly in the Soviet Union as well, over a half of total government-financed research and development is military research and development. For the world as a whole, about 40 per cent of research expenditure is devoted to military research.

About 400,000 of the world's most highly qualified physical scientists and engineers work on military research and development; this number is about 40 per cent of the world's research scientists and engineers. If only physicists and engineering scientists are included, the percentage is even greater—well over 50 per cent.

For the past two decades the bulk of military research and development has been done by the United States and the Soviet Union. These two countries account for about 85 per cent of the money spent on this activity. France, the United Kingdom, the Federal Republic of Germany and China together spent about 20 per cent as much as the United States. The rest of the world accounts for no more than 5 per cent of the total spent on military research and development. Of this group, the most significant spenders are Australia, Canada, India, Italy, Japan, The Netherlands and Sweden. The amount spent by Warsaw Treaty Organisation countries on military research and development is not publicly known. The countries spending the most on military research and development are, in general, also making the most rapid advances in military electronics.

Nuclear-war Fighting Weapons

The most dramatic consequences of advances in military electronics are those which make more probable a nuclear world war. Such advances include, in particular, improvements in the accuracy and reliability of strategic missiles, but also the development of space-based navigational aids, improvements in anti-submarine warfare techniques, the development of methods to destroy enemy ballistic missile warheads in space, and the development of ways of destroying enemy satellites in space.

Very accurate and reliable ballistic missiles are seen as suitable for fighting a nuclear war. For many years now, a large fraction—probably more than half—of strategic nuclear warheads has been targeted on military targets, even though these may usually have been large-area targets, often in or near cities. But with more accurate warheads smaller (and, therefore, many more) military targets can be targeted.

Even though many strategic nuclear weapons have been aimed at military targets, official nuclear policies were based on mutually assured destruction in which the enemy's cities were the hostages. The theory was that the enemy would not attack if his cities and industry could be destroyed in retaliation. Moves to a nuclear-war fighting strategy are being made not because the requirements of nuclear deterrence have changed (the psychology of the enemy is after all the same) but because military technology has made nuclear-war fighting weapons available. Once available, weapons are usually deployed. Policies then have to be modified so that politicians can justify this deployment.

The more the two great powers adapt to nuclear-war fighting doctrines, the greater the probability of a nuclear war will become, because the perception that such a war is "fightable and winnable" will rapidly gain ground.

Offensive and defensive strategic nuclear weapon systems may in due course be developed which will make a pre-emptive surprise attack possible, or, in the opinion of some scientists, probable or even inevitable.

Also considerably increased will be the outbreak of a nuclear world war by accident or miscalculation, either because of technological error or because of the misinterpretation of data.

The more the two superpowers rely on computers to warn them of nuclear attack and to launch and control their nuclear weapons, the greater the danger of accidental war becomes.

There have been several recent examples of false alarms being generated in American military computers. These false alarms were not caused by a real nuclear attack, but this was realised only after periods of up to 15 or 20 minutes. It is for this reason that the planned deployment of Pershing II missiles in the Federal Republic of Germany, targeted on locations in the European part of the Soviet Union, will considerably increase the danger of accidental nuclear war. The flight time of Pershing II to the Soviet Union is a mere 4 minutes or so. Given that Soviet computers are less sophisticated than their American counterparts, the danger for future accidents is obvious. Nuclear war is becoming increasingly hair-trigger.

Improving strategic missile accuracy

The accuracy of, for example, the current U.S. Minuteman III, the world's most sophisticated intercontinental ballistic missile (ICBM), is being upgraded by improvements made in the computer of the missile's guidance system. These involve better mathematical descriptions of the in-flight performance of the inertial platform and accelerometers, and better pre-launch calibration of the gyroscopes and accelerometers. With these guidance improvements the circular error probable (CEP)[1] of the Minuteman III will probably decrease from the current value of about 350 m to about 200 m.

Minuteman III ICBMs with the higher accuracy will be able to destroy Soviet ICBMs in silos hardened to about 4000 lb/in^2, with a

---

[1] For the explanation of these terms, see the Glossary of Technical Terms II at the end of the book.

probability of about 78 per cent for one shot and about 95 per cent for two shots.

The Americans are developing an exceedingly accurate new ICBM—the M-X—which may be virtually the ultimate in ballistic missile design.

The guidance for the M-X missile will probably be based on the advanced inertial reference sphere, an "all-attitude" system which can correct for movements of the missile along the ground before it is launched. A CEP of about 100 m should be achieved with this system. If the M-X warhead is provided with terminal guidance, using a laser or radar system to guide the warhead onto its target, CEPs of a few tens of metres may be possible.

An American strategic ICBM force of this accuracy would be seen by the Soviet Union as a considerable threat to its ICBM force. The most likely Soviet response to this threat would be the installation of a launch-on-warning system in which a computer would be used to launch Soviet ICBMs while the American missiles were in flight. A satellite system would be used to detect the American ICBMs as they crossed the horizon. A satellite signal would then trigger off the computerised launching procedures for the Soviet ICBMs. The initiation of a nuclear holocaust by computer, without any human decision, must surely be the ultimate madness. Even so, it is within sight.

Improvements of strategic submarine-launched ballistic missiles

The Soviet and American navies operate a total of 113 modern strategic nuclear submarines—the Soviet Union has 72 and the United States has 41. The ballistic missiles carried by these submarines are normally targeted on the adversary's cities to provide the assured destruction on which nuclear deterrence depends. A single modern U.S. strategic nuclear submarine, for example, carries about 200 nuclear warheads, enough to destroy every Soviet city with a population of more than 150,000. American cities are hostages to Soviet strategic nuclear submarines to a similar extent. Four strategic nuclear submarines (out of the 113) on appropriate stations could destroy most of the major cities in the Northern Hemisphere.

The quality of strategic nuclear submarines and the ballistic missiles they carry is being continuously improved. In the United States, for example, the present Polaris and Poseidon strategic nuclear submarine force is being augmented, and may eventually be replaced, by Trident submarines.

Trident submarines will be equipped with a new submarine-launched ballistic missile (SLBM), the Trident I, the successor of the Poseidon C-3 SLBM. Yet another SLBM, the Trident II, is currently being developed for eventual deployment on Trident submarines. In the meantime, Trident I missiles will also be deployed on Poseidon submarines.

The circular error probable (CEP) of the Trident I SLBM is probably about 500 m at maximum range, whereas that of the Poseidon SLBM is about 550 m. The development and deployment of mid-course guidance techniques for SLBMs and the more accurate navigation of missile submarines will steadily increase the accuracy of the missiles.

SLBM warheads may eventually be fitted with terminal guidance, using radar, a laser, or some other device to guide them onto their targets after re-entry into the earth's atmosphere. This could give CEP of a few tens of metres. SLBMs will then be so accurate as to cease to be only counter-city weapons and become nuclear-war fighting weapons able to threaten even relatively small military targets.

Soviet SLBMs are much less accurate than American ones. The SS-N-6 is thought to have a CEP of about 2000 m, the SS-N-8 a CEP of about 1300 m, and the SS-N-18 a CEP of about 1000 m. But we can expect that the accuracy of Soviet SLBMs will be steadily improved.

Anti-submarine warfare (ASW)

A very large effort is being put into improving ASW techniques by both the United States and the Soviet Union to detect and destroy enemy submarines. This will almost certainly lead, in time, to success. Even in the absence of a technological breakthrough—which cannot, of course, be discounted in spite of official reassurances—steady progress in limiting the damage that can be done by enemy strategic nuclear submarines will increase perceptions not only that a surprise attack may succeed but that it is essential.

In ASW, detection remains the critical element. Detection methods are being improved by increasing the sensitivity of sensors, improving the integration between various sensing systems, and better processing of data from sensors. Most current developments are, in other words, electronic.

All types of ASW sensors are being improved—electromagnetic ones, based on radar, infrared, lasers and optics; acoustic ones, including active and passive sonar; and magnetic ones, in which the magnetic field disturbance caused by a submarine is measured. Airborne, spaceborne, ocean-surface, and sea-bottom sensors are being increasingly integrated and, therefore, made more effective. ASW aircraft, surface ships, and hunter-killer submarines are also being made increasingly complementary. Each system has special characteristics and the integration of those that complement each other greatly enhances overall effectiveness.

A typical U.S. ASW task force would employ an ASW carrier equipped with specialised aircraft, destroyers equipped with ASW helicopters, and attack submarines. The task force would co-operate with land-based aircraft and receive information from unmanned surveillance systems—all to hunt down and destroy a single enemy strategic nuclear submarine. Once detected, the submarine would be relatively easy to destroy.

Soviet ASW is based mainly on naval helicopter carriers and long-range, land-based aircraft. Three Soviet helicopter carriers are in service, mainly intended for fleet defence. Search helicopters carry very sophisticated electronic equipment to detect and track enemy submarines. Armed helicopters destroy them. The Soviet Union also operates a number of ASW cruisers and destroyers.

Several types of Soviet long-range aircraft equipped with the most modern high-resolution radar and magnetic anomaly detection equipment are designed to detect and destroy U.S. strategic submarines. Soviet ASW activities tend to be much more short range than U.S. ASW activities and largely confined to areas close to Soviet territory or Soviet fleets.

The most effective single ASW weapon system is the hunter-killer submarine—a nuclear submarine equipped with sonar and other ASW sensors, underwater communications, a computer to analyse data from the sensors and to fire ASW weapons, such as torpedoes with active or passive acoustic terminal guidance and perhaps nuclear warheads. The United States and the Soviet Union each operate hunter-killer fleets, several dozen boats strong. A hunter-killer is designed to find and then follow an enemy strategic submarine until it is ordered to destroy it.

Cruise missiles

A strategic nuclear weapon system, the development of which depends almost entirely on microelectronics, is the modern cruise missile. This weapon is also extremely accurate and in the category of a nuclear-war fighting weapon.

Cruise missiles are old weapons, dating back to the German V-1 or "buzz-bomb" of World War II. Soon after the war, the United States and the Soviet Union began developing these missiles. A variety of types were produced (surface-to-surface, surface-to-air and air-to-surface) for both short-range (tactical) and long-range (strategic) applications.

In the early 1960s, U.S. long-range surface-to-surface cruise missiles were replaced by ballistic missiles.

In 1972 the U.S. interest in cruise missiles revived, according to some as a "bargaining chip" in the Strategic Arms Limitation Talks (SALT). A number of technological advances encouraged cruise missile development. The most important by far was the miniaturisation of computers in terms of volume and weight for a given power output. Also important was the availability of an accurate data base about the co-ordinates of potential targets. Very small but accurate missile guidance systems could thus be developed. For example, the McDonnell Douglas Terrain Contour Matching (TERCOM) system, which weighs only 37 kg, can guide a cruise missile to its target with a circular error probable (CEP) of a few tens of metres. TERCOM uses an on-board computer to compare the terrain below the missile (scanned with a radar altimeter) with a preprogrammed flight path. Deviations from the planned flight path are corrected automatically. From very accurate maps which have become available using satellite mapping techniques, the positions of targets and contours of flight paths can be obtained with unprecedented accuracy. Targets could not be located accurately enough from earlier maps to make effective use of the cruise missile guidance systems. These guidance systems are essentially based on pattern recognition, one of the main military uses of artificial intelligence.

Using these new technologies, cruise missiles are being developed in the United States to be launched from air, sea and ground platforms. Perhaps the most important characteristic of these cruise missiles is that the ratio of the payload carried to the physical weight of the missiles is relatively very high (typically about 15 per cent compared with a fraction of 1 per cent for a typical ballistic missile).

The air-launched cruise missile (ALCM), to be carried and launched from strategic bombers, has, for example, a very small radar image and flies at subsonic speeds at very low altitudes (a couple of hundred metres over rough terrain and a few tens of metres over smooth ground). The missile is difficult to detect and destroy. Defence against the missiles would thus be both difficult and costly, particularly if they were launched in large numbers. American strategic bombers will each carry up to twenty-four ALCMs equipped with nuclear warheads.

Effective detection of ALCMs would probably involve look-down radars carried in airborne warning and control system (AWACS) aircraft. To patrol a long frontier would require a fleet of such aircraft—a very costly undertaking. A large number of long-range interceptor aircraft would also be required to operate with AWACS to intercept and destroy the incoming missiles, which would also be extremely costly. The deployment of cruise missiles would, therefore, most probably escalate the arms race.

A defensive system against cruise missiles is generally more expensive than the cruise missiles themselves.

Despite their high accuracy and relative invulnerability, cruise missiles are quite cheap. In a production run of, say, 2000 missiles, the unit cost (including development costs) is likely to be about $750,000 (much less than the cost of a modern main battle tank).

Cruise missiles have considerable potential as strategic nuclear delivery systems for smaller countries. Britain and France, for example, are showing great interest in these missiles as potential cheap replacements for their strategic nuclear weapons as these become obsolete in the 1980s. Most industrialised countries (and possibly some Third World ones) are technically capable of producing cruise missiles indigenously. But what is often lacking is a precise knowledge of the co-ordinates of potential targets and accurate information about the flight path to navigate to their co-ordinates with the full effectiveness of the missile's guidance system.

If cruise missiles do proliferate widely, they may turn out to be the most far-reaching military technological development ever. And there is little doubt that improved versions of cruise missiles will, in future, be developed in rapid succession.

Airborne Warning and Control System (AWACS)

The AWACS is an exceedingly expensive piece of military high-technology, incorporating a large amount of microelectronics. The latest American AWACS aircraft, the E-3A, costs about $1.5 billion in research and development, and each aircraft costs about $100 million to produce, about six times the cost of the most sophisticated fighter.

According to current plans a fleet of 52 E-3As will eventually be built, 18 of which will be acquired by NATO. The aircraft will be used for the defence of North America, Central Europe, and the Greenland-United Kingdom gap. The Soviet Union operates a fleet of AWACS aircraft, based on the Tu-126 Moss, but Soviet AWACS aircraft are much less sophisticated than American AWACS aircraft.

Many regard AWACS as a microelectronic white elephant, in that the aircraft would be a prime and vulnerable target in a war, likely to be lost soon after the war begins. Also, of course, AWACS aircraft would have to operate in an environment of much electronic countermeasures (ECM). Critics claim that the radars are insufficiently effective against ECM.

AWACS enthusiasts argue that it can resist ECM better than current ground-based surveillance radars can, and that it would survive long enough in a hostile environment to fulfil a worthwhile role. They emphasise the value of an early warning period, however short, in a crisis.

AWACS aircraft have evolved from a line of airborne early warning aircraft. The first, the EC-121 Warning Star which became operational in 1951, was developed for high-altitude, long-endurance (about 20 hours at an altitude of 8 kilometres) radar surveillance at distances of up to 3500 kilometres from the American coasts. A crew of about 30 operated and maintained nearly five tons of radar and electronic equipment.

Then came the E-1 Tracer (operational in 1960) and the E-2 Hawkeye (1964). Hawkeye, designed to detect the approach of hostile supersonic aircraft soon enough to ensure interception, is equipped with highly sophisticated early warning search radars linked to a tracking and intercept computer which automatically provides digital target reports. A notable external feature of the aircraft is the 7.3 m diameter saucer-shaped rotodome which carries the radar aerial fixed above the fuselage.

A modern AWACS aircraft, like the E-3A, performs the functions of an airborne early warning aircraft and, in addition, provides extensive command and control facilities for all friendly aircraft within its range—including interceptors, transport aircraft, reconnaissance aircraft and so on. The identification and tracking of hostile aircraft and the control of friendly aircraft—at long range, at high and low altitudes, in all weathers, and over land or sea—are achieved by the use of highly sophisticated, beyond-the-horizon, look-down, high-pulse-frequency doppler surveillance radars, high-speed computers and multipurpose display units.

The E-3A is a modified Boeing 707. The liquid-cooled radar aerials are housed in a rotodome, hydraulically driven at 6 rpm, fixed over the fuselage. The aerials scan from ground level to stratospheric altitudes.

Because of the long patrolling times involved—up to about 11 hours, which may be extended to over 20 hours with air refueling—the E-3A has very accurate navigational systems. Typically, about a dozen AWACS specialists would be carried in each aircraft, in addition to the flight crew of four.

The Automated Battlefield

In tactical warfare the most concentrated application of microelectronics is in the development of the automated battlefield. In the past fifteen years or so considerable steps have been made in automating warfare—in ground, sea, and air combat. This is perhaps the most dynamic of current military technologies. It is also one in which there is a great deal of secrecy.

Most of the automatic systems being developed for the battlefield rely much on microelectronics. The development of new electronic offensive weapons inevitably stimulates the development of electronic countermeasures against them. This in turn provokes the development of electronic counter-countermeasures, and so on. It is hardly surprising that military microelectronic revolutions follow one another with bewildering rapidity. In this field no one individual can hope to keep abreast of all the developments, even if he could get at the information.

Most battles have four distinct phases. Firstly, the enemy forces are located and identified. Secondly, a decision is made about how to deal with these enemy forces. Thirdly, appropriate weapons are fired on the enemy. Finally, the damage done to the enemy is assessed to find out if the sequence needs repeating.

In the automated battlefield the enemy forces—men and vehicles—would be detected by remotely piloted vehicles or sensors planted on the ground. The data so collected would be transmitted back to central computers. These computers would decide on the action to be taken and then direct the weapons onto their targets. After the weapons had been fired, sensors on the battlefield and remotely piloted vehicles would assess the damage done, feed the information back to computers which would decide whether or not to fire more weapons, and so on.

Currently, many of the automatic systems necessary for an automated battlefield are available and in use. Others are being developed. So much information on this subject is classified that it is difficult to assess just how close the fully automated battlefield is. But, given the state of the art of microelectronics we can be sure that the United States is well ahead of other countries in automated battlefield techniques.

General William Westmoreland, the ex-Chief of Staff of the U.S. Army, has predicted that "on the battlefield of the future, enemy forces will be located, tracked and targeted almost instantaneously through the use of data links, computer-assisted intelligence evaluation, and automated fire control. With first-round kill probabilities approaching certainty, and with surveillance devices that can continually track the enemy, the need for large forces to fix the opposition physically will be less important''. From the available evidence it seems likely that the automated battlefield envisaged by the General will become a reality in the 1980s. Microelectronics will play the essential role in this development.

Sensors

The sensors on the automated battlefield can be sensitive to light, sound, electromagnetic waves, infrared radiation, magnetic fields, pressure, chemicals, and so on. Of most interest here are sensors which transmit information about the enemy forces and movements by radio over long distances.

The most common types of sensors so far developed for the battlefield are those which pick up seismic disturbances in the ground caused by the movements of people and vehicles. They can be implanted on the battlefield by hand or dropped by aircraft. They are usually buried in the ground with just an aerial visible. People can be detected by typical seismic sensors at distances of up to 50 metres or so and vehicles at ten times this distance. The lifetime of the sensors is normally a few months.

Seismic sensors are often used with acoustic ones. A seismic sensor cannot normally distinguish between a light vehicle at a close distance and a heavier one further away. If an acoustic sensor is used to monitor the same target, a more certain identification can be made. In fact, it would be normal to use many types of sensors in combination to be certain of detecting a high fraction of enemy targets.

A main role of microelectronics in sensors is to make their size small even though the power supply of their transmitters is strong enough to give ranges of a few kilometres or more. As a rule, ground relay stations would, when necessary, be used to transmit sensor signals over very large distances to the central computers. Relay equipment may also be carried in high-flying aircraft, which may be remotely controlled.

Human operators sitting in control stations away from, or flying over, the battlefield would access the data from the sensors and direct appropriate weapons to the targets. But this phase of the battle may, in the future, also be automated. A computer could be programmed to react to information from sensors and issue orders for further reconnaissance or direct attack.

Stand-off weapons

The weapons most usually thought of for use on the automated battlefield are guided weapons—surface-to-surface missiles, air-to-surface missiles, and guided bombs. These may be fitted with automatic homing devices so that, once launched, the missiles will seek out the target without further external help.

A typical stand-off weapon has an electro-optical remote guidance and control system and a solid-propellant rocket engine giving the missile a range of about 100 kilometres. As the missile travels to its target area, after launch from the parent aircraft, it returns data-link television pictures to an operator in the parent aircraft who has a display unit on which the terrain in sight of the TV camera mounted in the nose of the missile is displayed. The operator can change the missile's course, select a target, and then lock the warhead onto the target by sending radio command signals to the missile over the data link.

These stand-off missiles will probably give way to a new generation of remotely-piloted vehicles (RPVs) built for ground attack missions, with ranges of 200 kilometres or so. RPVs will also be used for reconnaissance, electronic warfare, and even air-to-air combat. The development of these small RPVs is made possible by microelectronic systems for guidance, control and communications. Using TV cameras and data transmission links, and so on, an RPV could be controlled precisely from a safe distance from the battlefield, either by a pilot in a launch aircraft or by operators on the ground using TV pictures transmitted from the RPV.

A typical RPV of the type now under development will carry about 500 kilograms of cameras, electronic intelligence receivers, side-looking radar, communications relays, etc., will operate at altitudes of up to 25,000 metres or so, have speeds of about 500 kilometres per hour, and take off and land on a runway. For large-scale tactical warfare a very large number of such RPVs would be used and controlled through computerised command and control centres. The vehicles are cheap enough to make such use in large numbers possible.

Modern Avionics

A modern combat aircraft is an excellent example of a weapon system whose performance relies on the most up-to-date electronics. The most advanced developments in avionics are to be found in air superiority fighters, aircraft designed to seek, find and destroy any type of enemy aircraft, whatever the weather. Such an aircraft may also be capable of other missions, like air-to-surface attack.

A modern combat aircraft's avionics typically includes a lightweight radar system so that the aircraft can detect and track high-speed targets at great distances and at all altitudes down to tree-top level. The tracking information is normally fed into the aircraft's central computer for the accurate launching of missiles or the firing of an internal gun. For close-range combat, the radar automatically projects the target on a head-up display which details all the necessary information, as symbols, on a glass screen positioned at the pilot's eye level. The pilot can thus be automatically provided with the data required to intercept and destroy an enemy aircraft without taking his eyes off the target. The head-up display also provides the pilot with navigation and control information under all conditions and with details about the aircraft's performance so that he can detect any faults which develop in any of the aircraft's systems as soon as they occur.

An "identification-friend-or-foe" system informs the pilot if an aircraft which he has detected by eye or radar is a friend or foe. And an air data computer and an attitude and heading reference set display information on the pitch, role and magnetic heading of the aircraft. This, together with an inertial navigation set, enables the pilot to navigate anywhere in the world.

The United States is considering a so-called quick-reaction interceptor aircraft, having speeds greater than Mach 3 (three times the speed of sound) and extremely rapid acceleration, for operations with future air-defence weapons systems. Such an aircraft may have air-superiority and tactical strike roles in addition to the intercept role. Even faster aircraft with hypersonic (up to Mach 10) speeds are also under active consideration. It is mainly advanced avionics which make such extraordinary combat aircraft feasible.

A typical air superiority fighter would be armed with air-to-air missiles of short and medium ranges. The U.S. Sparrow is an example of an all-weather, all-aspect, air-to-air missile. It uses a continuous wave semi-active radar guidance system and carries a proximity-fused high-explosive warhead weighing about 5 kilograms. A solid propellant rocket propels the missile at a speed of Mach 3 over a range of about 50 kilometres.

The avionics in the missile is miniaturised: the consequent reduction in the volume of avionics allows a larger solid-propellant motor to be installed. The missile also has good electronic counter-measure (ECM) capabilities, is very reliable and manoeuvrable for short-range combat.

The very latest Sparrow air-to-air missiles combine radar and infrared in their target-seeking system. Missiles with semi-active radar guidance require the target to be continuously illuminated. The aircraft's radar transmits continuous-wave illuminating signals so that, after launch, the missile can home on the reflections of these signals from the target. Because the aircraft's radar antenna has to be directed at the target until the missile hits it, only one target can be engaged at a time. But heat-seeking missiles carry their own terminal guidance and, therefore, need only be launched in the direction of the target.

Electronic Warfare (EW)

Continuous improvements in defence radars, anti-aircraft artillery systems, surface-to-air missiles and interceptor aircraft stimulate efforts to develop electronic counter-measures designed to frustrate the operation of the enemy's electronic devices associated with all of his defensive systems. This, in turn, stimulates the development of electronic counter-counter-measures (ECCM) to prevent interferences by electronic counter-measures (ECM).

A whole range of electronic warfare (EW) equipment has been developed to gather and co-ordinate as much data as possible about the adversary's radar (land, sea, air, surveillance, weapon control, navigation, etc.), command and control, and communication systems, a form of EW known as Elint (electronic intelligence). Effective Elint requires worldwide land, sea, air and outer space information-gathering operations, an enormous and very costly enterprise, using the most advanced electronic equipment. The knowledge thus gained of the potential enemy's electronic equipment is then used to develop appropriate ECM and ECCM devices and procedures.

The Superpowers' Arms Race

Microelectronics has become the main factor in the technological arms race between the two superpowers. By continuous effort in military technological development, the Americans are striving to retain a technological superiority in as many weapons systems as possible, a superiority which the Soviets are seeking to eliminate. A fundamental American belief is that, provided sufficient resources are devoted to research, the nature of their political system is such as to encourage innovation more than other political systems do. This then is a field of activity, and possibly the only one, in which the United States hopes to be able to keep permanently ahead of the rest of the world. Failure would, so it is believed, bring military or political disaster. But such opinions, based on perceptions rather than facts, are by no means confined to Americans. Political leaders in general appear to believe that military technological superiority prevents potential adversaries from applying pressures and blackmail in international affairs, or brings victory in war.

The struggle by both the United States and the Soviet Union for military pre-eminence is by far the most important single cause of the on-going arms race (nuclear and conventional) between them. The arms race will undoubtedly continue until both sides cease attaching such enormous political importance to gaining and retaining technological superiority. Past arms races have led to wars; the remainder have led to economic collapse. There is no reason to believe that modern arms races are exceptions.

Although the most dramatic advances in weaponry have occurred in strategic arms, continual improvements are being made in the quality of virtually all tactical weapons. In very many cases, the greatest advances come from microelectronics developments.

Western official assessments credit existing Soviet deployed forces with technological superiority in anti-ballistic missile systems, strategic air-defence interceptors, all aspects of civil and industrial strategic defence and recuperative planning, tactical anti-ship missiles, surface attack ships (excluding carriers), anti-aircraft artillery systems, some armoured combat vehicles, medium- and high-altitude surface-to-air missile air defences, surface-to-surface tactical missiles, and heavy-lift helicopters. Approximate technological parity is said to exist in deployed systems such as tanks and anti-tank weapons, satellite tracking systems, satellite navigation systems, and small arms. But the Americans still claim to retain a technological lead in intercontinental ballistic missile guidance and penetration aids, strategic bombers, strategic submarines and submarine-launched ballistic missiles, attack submarines, anti-submarine warfare sensors and patrol aircraft, satellite communications systems, airborne warning and control systems, airborne surveillance sensors, defence-suppression weapons and systems, deep-strike tactical aircraft, aircraft carriers, guided ordnance, air-to-air superiority weapons, man-portable air-defence systems, close-support helicopters, aircraft and aerial weapons, long-range logistic transports, and artillery. This list clearly reflects American superiority in all aspects of microelectronics.

The dynamics of superpower military technology can be illustrated by one example, the military use of high-energy lasers. Many others could be given. The rapid progress in production of high-energy lasers (continuous-wave power outputs of a few hundred kilowatts can now be achieved) has stimulated great interest in the development of thermal laser weapons. Likely applications for the first generation of these revolutionary weapons, in combination with appropriate electronics, include ground-based air defence against low-flying aircraft, missiles and remotely-piloted vehicles (RPVs) (by the thermal destruction of critical parts of the attacking vehicle), and air-to-air combat. The range of military applications of high-energy lasers will undoubtedly increase as their power output increases, including, perhaps the development of a space-based defence system against enemy ballistic missile warheads.

Worldwide Military Communications

Because of the great effort they have put into military research and development, the superpowers have developed military machines far outstripping those of other powers. There is even an enormous gap between the United States and the Soviet Union, on the one hand, and their nearest rivals, such as the French, British and West Germans, on the other hand. This is dramatically illustrated by, for example, the U.S. Worldwide Military Command and Control System. This is based on a vast network of information gathering systems, on land, sea, in the air, and in space. No smaller power could hope to operate such a gigantic undertaking. The United States spends in excess of $1000 million a year on the system which employs about 90,000 people just to operate communications and the command centres.

The purpose of the system is to enable the U.S. National Command Authority (the President, the Secretary of Defense, and authorised successors) to have operational control of the strategic forces at all levels of combat. This includes providing the means by which the National Command Authority can receive warning and intelligence of enemy actions in order to make its decisions, assign appropriate military missions and direct the various military commands.

At the centre of the worldwide system is the National Military Command System, which includes the command centres and the communications used by the National Command Authority. The main command centre is in the Pentagon, but there is an alternate centre near Fort Richie and an airborne command post. In addition to this command system, four American Commanders-in-Chief (of the Strategic Air Command and of the European, Atlantic, and Pacific areas) have fixed and airborne command posts capable of communicating worldwide with the nuclear forces. The Commander-in-Chief of the Strategic Air Command keeps his command post continuously airborne.

The airborne command post is designed to allow the National Command Authority staff to fight a nuclear war even after the fixed ground-based command centres and communications networks have been destroyed. The aircraft used are Boeing 747s modified to provide a conference room, a briefing room, a battle staff work area, and a communications control centre. Three of these aircraft are in operation to provide airborne command, control and communications for the National Command Authority and the Commander-in-Chief of the Strategic Air Command. Three more are to be procured. The fleet is being modernised, in particular with new super high frequency and very low frequency communications systems, and automatic data processing equipment.

There are direct communications between the command aircraft, on the one hand, and the intercontinental ballistic missile wings and the aircraft relaying messages to strategic nuclear submarines, on the other hand.

Worldwide communications links to strategic nuclear forces and tactical nuclear weapon storage sites are provided by a special satellite communications system. The system consists of ultra-high frequency communications transponders on several communications satellites. Satellite terminals are installed on airborne and ground-based command posts, strategic bombers, reconnaissance aircraft, intercontinental ballistic missile launch control centres, etc.

The U.S. Navy keeps an aircraft continuously airborne over the Atlantic to ensure that National Command Authority orders can be relayed to strategic nuclear submarines in the area, even if fixed ground-based transmitters have been destroyed in a nuclear war. A similar continuous airborne system will soon operate over the Pacific. The aircraft are equipped with powerful very low frequency transmitters for communication with submarines.

An early warning satellite system and ground-based radar systems are designed to provide the National Command Authority with warning of attack by intercontinental ballistic missiles or submarine-launched ballistic missiles. Ballistic Missile Early Warning System radar sites are operating in Greenland, Alaska, and England. Perimeter Acquisition Radars are in operation in the United States. Surveillance radars to confirm satellite warning of submarine-launched ballistic missile attack are deployed along the American coast. These systems are being continuously improved as new electronics become available, in particular by the installation of new computers and new phased-array radars.

Over-the-Horizon-Backscatter radar is being tested to provide long-range surveillance of aircraft and warning of a bomber attack. And the Distant Early Warning Line radars are being improved, particularly to detect aircraft flying at low altitudes.

The Integrated Operational Nuclear Detection System is being developed to increase U.S. capability to detect, quickly locate, and report nuclear explosions on a worldwide basis. The system will provide nuclear trans- and post-attack damage assessment information for the National Command Authority. Current nuclear detection sensors are carried on the early warning satellites.

The American worldwide military communications system involves a network of bases, including stations in North West Cape, Australia, Diego Garcia in the Indian Ocean, Guam in the Pacific, etc. There are also worldwide networks of radars—like the U.S. Air Force Satellite Control System to track, control, and interrogate military satellites.

In addition to all of this, there are command, control, communications, and intelligence systems for tactical purposes supporting land, sea, marine, and air forces, as well as theatre nuclear forces. These include a network of radar stations across NATO countries to detect enemy aircraft, ground mobile forces satellite communications involving hundreds of the various types of transportable terminals, ultra-high-frequency satellite communications with theatre nuclear forces, and so on.

The Defense Satellite Communications System, a super-high-frequency satellite communications system using seven satellites, links the continental United States with military forces abroad. Large fixed and mobile terminals overseas link in with the Worldwide Military Command and Control System and tactical systems.

The Defence Communication System provides American military forces with worldwide voice, data, and teleprinter services. This is being made more flexible and interoperable with the systems of America's allies. New electronics are continuously being used to make the main radio links more secure from jamming, interception, electronic eavesdropping, and interference.

Given the demands made on modern military communications, it is hardly surprising that the superpowers are investing large sums in the field. The U.S. Department of Defense, for example, is spending more than $3000 million a year on computer software. Much effort is going into qualitatively improving the production of software. There is also an active multi-million dollar programme to accelerate the introduction of advanced integrated circuit technology into military systems, and relating this to the problem of interoperability and software.

Also under continual and rapid development are technologies for survivable computer communications, secure message and information-transfer systems, crisis management and command systems, and miniature computers.

The money and skills involved in large-scale military command, control, communications, and intelligence operations are clearly extremely big. Only superpowers can readily afford the effort.

Military Electronics Outside the Superpowers

Any country with a significant defence industry producing aircraft and missiles is most likely to develop also a military electronics industry, particularly if it intends to export weapons.

Industrialised countries, other than the United States and the Soviet Union, producing military aircraft include: Australia, Belgium, Canada, Czechoslovakia, Finland, France, the Federal Republic of Germany, Italy, Japan, The Netherlands, Poland, Romania, Spain, Sweden, Switzerland, the United Kingdom, and Yugoslavia. Third World countries producing military aircraft include: Argentina, Brazil, China, Egypt, India, Indonesia, Israel, North Korea, South Korea, Pakistan, the Philippines, South Africa, and Taiwan. Except for Czechoslovakia, Finland, The Netherlands, Poland, Romania, Spain, Switzerland, Indonesia, North Korea, South Korea, and the Philippines, all these countries also produce missiles.

Of the Third World countries listed, Argentina, Brazil, China, India, Israel, and South Africa have significant defence industries and participate in the global arms trade. By and large, these Third World countries and the industrialised weapons-producers are developing modern military electronics systems. But none can compete with the United States and the Soviet Union in the range and sophistication of equipment, nor can any keep up with the superpowers in research and development into military electronics.

Most Third World countries and many industrialised ones acquire their most sophisticated weapons through the global arms trade. An impression of the extent of this trade can be had from the fact that about $120,000 million a year is spent on procuring weapons: more than $30,000 million worth of these weapons are acquired from other countries.

In many cases the countries buying sophisticated weapons are unable to operate them effectively or maintain them properly. This is particularly true for those weapons relying on advanced electronics.

About forty Third World countries operate, for example, modern combat aircraft which rely for efficient operation on sophisticated avionics. Many of these countries are unable to handle this equipment effectively. An extreme example of this was the attempt by Iran before the revolution to operate E-3 Airborne Warning and Control System (AWACS) aircraft.

The fact that only the superpowers can produce the equipment to automate warfare and to operate the most up-to-date real-time command, control, communications, and intelligence systems means that some types of warfare can now only be waged by these powers. The enormous superiority that the superpowers have because of their high military technology gives them considerable political influence in many instances. For example, one American aircraft carrier packs a punch comparable with that of many countries' entire military forces, even excluding the nuclear weapons carried by the aircraft carrier. The appearance of such a ship off a smaller country's coast obviously can influence the actions of that country.

The gulf between the military technological capabilities of the advanced countries compared with the others is steadily increasing. This has obvious consequences for regional and world power balances.

An example of the increasing gulf in tactical warfare is navigation. The United States is now deploying a satellite navigational system called Navstar, or the Global Positioning System, due to be operational in 1984. This is based on twenty-four satellites and will provide navigation capability for ground-based, airborne or shipborne weapon systems of unprecedented accuracy. A navigator will be able to obtain continuous position fixes in three dimensions to within 10 metres or so, and will be able to calculate his velocity to within a few centimetres per second. The system will enable missiles to be accurately navigated to their targets and will help synchronise the automated battlefield. A country which has access to such a system will have a considerable military advantage over one which does not.

There are, of course, many who question whether the extraordinarily advanced technology being deployed by the military of the great powers will work effectively under battlefield conditions. Most of the middle-ranking officers now planning military tactics are, it is pointed out, often technological enthusiasts with little or no real experience of war. They are inclined to introduce technological innovations purely for the sake of having the most up-to-date equipment with scant regard to its practicability under fire. It is significant that the generals who have had battle experience, now retired or with diminishing influence on the course of events, are often among the critics of the amount of high technology being introduced into weapon systems.

Defensive Deterrence

Because of their dependence on advanced technology, modern offensive weapons are exceedingly expensive. Each generation is considerably more expensive than the last. Consequently, only the larger powers can readily afford the costs of offensive weapons. This trend may, in time, force the smaller powers to adopt a defensive deterrence strategy based on relatively cheap anti-tank and anti-aircraft missiles. The gap between the costs of offensive and defensive weapons is already large and is growing fast. A modern combat aircraft costs about $60 million whereas the cost of ground-to-air anti-aircraft missiles range from about $10,000 to $500,000. A modern cruiser costing about $200 million can be destroyed by an anti-ship missile costing about $500,000. And a battle tank costing $1,500,000 can be destroyed by an anti-tank missile costing less than $10,000. A policy based on defensive weapons may be combined with total civilian defence. The object would be to make the military occupation of the country so expensive that a potential aggressor would be deterred from attacking in the first place.

Such an approach to national defence has already been adopted by Sweden, Switzerland, Yugoslavia and Romania. The division of the world between the big powers with massive offensive capabilities and the rest with only defensive ones would bring new security problems, as well as solving others. For one thing, the credibility of defensive deterrence may be questioned. Moreover, the stronger powers may be able to apply political pressures on the others by, for example, making threatening moves with their offensive forces.

Nevertheless, defensive policies based on conventional rather than nuclear weapons may become important for some countries—even those in the alliances, both in NATO and the Warsaw Treaty Organisation. The deployment of new nuclear-war fighting weapons in Europe will probably provoke considerable public opposition to nuclear weapons and demands for their removal from some countries' territory. Non-nuclear defence policies may then become more popular, and politically and economically attractive.

Microelectronics would play a fundamental role in such a defence policy—ranging from use in electronic barriers around frontiers to use in reconnaissance systems and rapid communication systems for small, highly-mobile, groups of soldiers armed with defensive missiles which can deal with offensive hostile weapon systems which manage to penetrate into the defended territory.

Military Science and the Increasing Probability of Nuclear World War

Even though the ever-increasing gulf between the military power of the United States and the Soviet Union as compared with all other countries has serious consequences for world security, it is overshadowed by the increasing probability of nuclear world war brought about by developments within the superpowers, mainly military technological developments.

The world's arsenals contain tens of thousands of nuclear weapons, probably topping 60,000. The total explosive power of these weapons is equivalent to about one and a quarter million Hiroshima bombs, or about 4 tons of TNT for every man, woman, and child on earth. If all, or a significant portion of them, were used, the consequences would be beyond imagination.

All the major cities in the Northern Hemisphere, where most nuclear warheads are aimed, would be destroyed (on average, *each* is targeted by the equivalent of 2000 Hiroshima bombs). Most of the urban population there would be killed by blast and fire, the rural population by radiation from fallout. Many millions of people in the Southern Hemisphere would be killed by radiation. And the disaster would not end even there. The unpredictable (and, therefore, normally ignored) long-term effects might well include changes in the global climate, severe genetic damage and depletion of the ozone layer that protects life on earth from excessive ultraviolet radiation. No scientist can convincingly assure us that human life would survive a nuclear world war.

Utterly catastrophic though a nuclear world war would be, its probability is steadily increasing because military scientists are developing weapons which will be seen as suitable for *fighting* rather than *deterring* a nuclear war. As we have seen, these new weapons include very accurate and reliable ballistic missiles with warheads that can be aimed at smaller and, therefore, many more military targets than in the past. In other words, the day is coming when one country might hope to destroy its enemy's nuclear retaliatory capability by striking first.

In this context, a first strike does not mean the ability to destroy totally the other side's strategic nuclear forces in a surprise attack. What it does mean is that the attacker *perceives* that he can destroy enough of the enemy's retaliatory forces in a surprise attack to reduce the casualties he receives in a retaliatory strike to a number he regards as "acceptable" for a given political goal. (In nuclear issues, as in almost all matters of international politics, perceptions determine events; facts do not.) The more reckless the political leadership is and the more tense superpower relations are, the greater this number of casualties is likely to be. And in his calculations the attacker is likely to make assumptions about the performance of his own and the enemy's nuclear forces which suit his arguments. Specifically, military calculations are likely to be based mainly on estimates of prompt deaths and ignore the uncertain long-term effects of a nuclear war which may well be far more lethal than the early effects. Also, in times of crisis political leaders listen to their military chiefs rather than their scientific advisors.

The direct attack by one superpower on the other may not be the most likely way in which a nuclear war will start. A more likely way may be the escalation of a future conflict in a Third World region. This may begin as a conventional war and then escalate to a nuclear war using the nuclear weapons of the local powers. This could escalate in turn to an all-out nuclear war between the superpowers.

This escalation is most likely if the superpowers are already involved in the Third World conflict, because they supplied the conventional weapons for the original conflict. Modern war uses munitions, particularly missiles, so rapidly that Third World countries at war need new supplies of weapons very rapidly. In this way, the arms supplier virtually guarantees the survival of its client at war. This involves the superpowers in regional conflicts and carries with it the risks of escalation to a total nuclear war. It is because of its contribution to the threat of the escalation of a third world war that the international arms trade is so dangerous.

The superpowers use these conflicts to test their most sophisticated weapons. The massive use of the most advanced microelectronics in modern weapons has much influence on this situation. It increases the demand for sophisticated weapons by Third World countries so that they can have the most up-to-date—and, it is assumed, the most effective—weapons. It increases the temptation of the superpowers to test continually the electronics in their missiles and other weapons by their use in Third World conflicts. It increases the violence of wars. And it increases the probability of escalation of conflicts.

The superpowers can be blamed for the danger of nuclear catastrophe which we all face. They, and their main allies, after all, spend the bulk—about 80 per cent—of world military expenditure, and own the vast majority of the nuclear warheads in the world today (Fig. 1). The Third World spends relatively little—about 13 per cent—of total world military expenditure. The superpowers also supply most—about 75 per cent—of the arms transferred to the Third World (Fig. 2). Nevertheless, most conflicts take place in the Third World, and there have been about 140 conflicts since World War II. And any future Third World conflict may escalate to a nuclear world war.

Improvements in warhead design and missile accuracy, which have virtually reached their theoretical limits, are just two examples

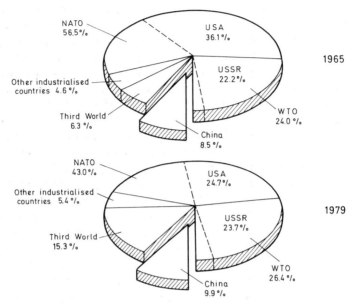

Fig. 1. Distribution of world military expenditure 1965 and 1979. Source: SIPRI Yearbook 1980.

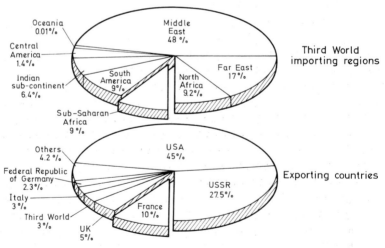

Fig. 2. The importers' and exporters' shares of major-weapon supplies to the Third World, 1970-79. Source: SIPRI Yearbook 1980.

which indicate the incredible progress made by military technology since World War II. In the next thirty years we can expect many more military technological revolutions. Many of these developments—which more often than not will be based on advances in microelectronics—will contribute to perceptions that a nuclear war is fightable and winnable, and that a first strike is feasible and even essential, on the argument that "Unless we strike now the other side will soon do so."

An excellent example is anti-submarine warfare. Now that land-based intercontinental ballistic missiles are, or soon will be, vulnerable to a first strike by missiles, nuclear deterrence depends on the invulnerability of submarine-based ballistic missiles. The fact that the superpowers continue working so energetically on anti-submarine warfare shows that they are unable to control military science and technology even though this activity is jeopardising their policy of nuclear deterrence, a policy the leaderships desperately want to maintain.

Perhaps one should emphasise here the military use of space. The application of microelectronics is the main reason for, and makes possible, the rapidly increasing militarisation of space. Particularly disturbing is the development and deployment of space-based, first-strike technologies. These include, in addition to reconnaissance satellites (spies in space), early warning systems, navigational systems, command, control, communication and intelligence systems, anti-satellite warfare systems, and ballistic missile defence systems. Since the space age began in 1957, about 1200 satellites have been launched. About 75 per cent of them have been for military purposes.

The large group of scientists who rely entirely on military money for support is, of course, a powerful political lobby. Moreover, vast bureaucracies have grown up in the great powers to deal with military matters. (As many civilians are paid out of military budgets as there are troops in uniform.) Academics and bureaucrats join with the military and defence industries to form an academic-bureaucratic-military-industrial complex intent on maintaining and increasing military budgets and agitating for the use of every conceivable technological advance for military purposes. This complex has so much political power as to be almost politically irresistable. In fact, the nuclear arms race is now totally out of the control of political leaders. *And this is as true in the Soviet Union as it is in the United States.* The uncontrolled nuclear arms race is without doubt the greatest single threat to our survival.

This is not to deny that great efforts have been made to control military technology and to stop the nuclear arms race between the Americans and the Soviets. Since World War II many of the world's most brilliant people have been actively involved in these efforts. No other problem has received so much attention in the United Nations and other international forums. Whole libraries have been written on the subject. Yet, because of the enormous political influence of those groups that continually press for greater military efforts, nuclear and other arms races go on just as fast as human skill in the American and Soviet societies allows. We are being driven toward nuclear world war by the sheer momentum of military technology.

I am not suggesting that some evil group is plotting our destruction. But I do suggest that we may not have the sort of intelligence required to set up the political and social institutions essential to controlling military technology. So far as I can see, we are drifting toward disaster not because politicians are either ignorant or want this to happen, but because of man's very nature.

The main, perhaps the only, hope for the future is that the public will learn the facts in time and that an aroused public opinion will force reluctant politicians to stop the nuclear arms race and reduce armaments. I am convinced that political leaders, left to themselves, will not be able to prevent a nuclear holocaust, even though they may sincerely wish to do so. I am equally convinced that if the public knew the truth about the nuclear arms race, it would insist on action by its political leaders to stop this insanity. We will avoid nuclear disaster only if the public protests in time.

9

Information Technology and Society

KLAUS LENK

Information Technology and Microelectronics

Unlike the previous chapters of this book, the following discussion concentrates not on microelectronics but on information technology. By "information technology" we understand technologies relevant to human communication processes and to the handling of what is conveyed in these processes—i.e. information. Applications of microelectronics in end-products like watches, radio sets, etc., as well as in the production process are therefore neglected.

Information technology and microelectronics are largely overlapping. Information technology englobes such different things as bookprint, reprography, the telephone network, broadcasting, the typewriter and the computer. Many of them are candidates for the building in of some amount of microelectronics. There is one case, however, where microelectronics is so interwoven with a particular information technology that they are almost identified in public opinion. It is precisely this case of computing, automated data processing (ADP), which is of outstanding interest in our context. Another area where microelectronics is very important is communications technology, including not only mass media but also interactive media like the telephone, or computer/communication networks.

Concentrating on the societal implications of information technology should of course not make us forget that the building in of microelectronics in end-products, as well as its use in the production process, can have important societal implications as well. Like other technologies, microelectronics can be of help in man's struggle to

273

dominate nature. And like the technologies of the Industrial Revolution it can heavily impact on society by the working conditions it imposes and by the products it helps to create.

Yet an important difference remains. The present explosion of information technology and of microelectronics is much more closely related to the functioning of society as a whole than was the Industrial Revolution. To a much greater extent than other technologies, microelectronics affects the very essence of social cohesion, i.e. communication. Information and communication constitute the fabric of society in more than a metaphorical sense. They do not remain unaffected when communication processes are mediated, channelled and partly taken over by technical devices.

There are at least four characteristics of information technology which in one way or another affect society by shaping its information and communication processes. Information technology can first provide a storage medium. The changeover from oral to written tradition is probably the oldest societal impact of information technology. Later, with the telegraph and the telephone, information technology extended its capability from bridging time to bridging space. Thus, the second important characteristic of information technology for society is the potential of telecommunications to extend the available range of man's vicarious experience, of eventually making the world shrink into one global village.

Both characteristics bring about effects that are not new to society. Nor are they uncommon. In his book *The Bias of Communication,* Harold Innis marshals pervasive evidence for the societal impact of many classical and new information technologies. To give but one example, the secularisation of European thinking is in his view closely linked to the spreading of bookprint.

The two remaining characteristics of information technology could be neglected until the advent of the computer. One is the automation of control, the other being the mechanisation of arithmetic and logic.

Automation of control is not confined to information technology in the sense defined above. It is of still more importance in production processes than in human communication processes. However, automatic monitoring of patients in hospitals or of traffic at intersections provides examples where society and the individual are affected.

Most important in our context is the other mentioned characteristic of computers, namely their ability to perform formalised operations. Computers are more than just calculating machines. This makes them candidates to take over formalised (mathematical and logical) thinking from human beings. In many ways, computers combine this function with the functions of information storage and transmission, and of automation of control.

It is precisely this combination of functions which has led to most of the existing computer applications connected to societal information and communication processes. Computers serve as tools for the collection, storage, manipulation and retrieval of very large information pools. In most of the existing administrative applications they serve as fast typewriters and desk calculators, or as media for the storage and retrieval of formatted information. To this can be added applications of computers as parts of communications systems and as monitoring or steering devices.

Most of these applications make only limited use of the distinctive capacity of computers, namely their ability to give answers to questions by means of computation and logical inference.

The Implications of Computing in Organisations

At present, the most tangible societal implications of information technology emanate from automated data processing (ADP) supported information systems and processes in both private and public organisations. Information systems support many internal and external functions of organisations in business and government. Personnel management, accounting, banking and insurance operations, social security administration, statistics and population registration are but a few examples.

The penetration of both public and private administrations by ADP differs widely from country to country, even among highly industrialised countries. A closer look soon reveals that the pace of change is dictated less by the available technology than by cultural factors including the policies of decision makers in charge of promoting new ADP systems. Such policies were in the past quite often tinged by exaggerated expectations with regard to what computers can do. Better control by governments, by top administrators or by corporate executives of the operations of sub-units seemed to come within reach as a by-product of massive computer deployment. Many policy makers expected sweeping changes to take place in the way bureaucratic organisations do their work, organise their contacts with their public as well as with their own staff.

Yet the present situation of ADP in public administration is slightly different from these expectations, and so is the situation in business administration. In either case this is not equivalent to saying that there are no impacts at all. Only quite recently have there been serious empirical studies of such impacts. They deal mainly with the role of computer use in decision making in organisations, the work life of computer users, and the alteration in patterns of power and managerial control.[1] Perhaps the major question raised in this research refers to the extent to which computers may contribute to the centralisation of control, i.e. the movement of decision authority up in the chain of command.

Other expected impacts like alterations of the size and formal structure of organisations are not yet supported by conclusive evidence. Organisations which have first computerised much of their routine work include banks, insurance companies as well as taxation authorities and social security agencies. Research undertaken in insurance companies indicates that these companies centralised their offices when they automated. The locus for decision making moved upward in the organisational hierarchy. Clerical jobs often diminished in scope, variety and autonomy. To many white-collar workers, computing appeared as an external force to which there was no choice but to adapt.

In the past, the introduction and use of complex and centralised ADP systems for the execution of administrative routine tasks frequently went hand in hand with a more elaborate division of labour. There was a tendency to split up continuous processes formerly carried out by one and the same public official. Those parts of such

---

[1] For a critical overview see Rob Kling, Social analyses of computing: theoretical perspectives in recent empirical research, in *Computing Surveys,* vol. 12 (1980), pp. 61-110.

processes that could not be consigned to machines were assigned to different persons. In short, Tayloristic forms of division of labour have invaded clerical work at the same time as they are losing ground in the production sector.

This not only entails an evaporation of the responsibility of the single agent who is no longer in charge of the full decision. It may also bring about changes in the decision-demanding content of individual work roles. This would be in direct contrast to job enrichment strategies. The autonomy and scope of action of organisation members may be eroded by ADP systems which generally limit their discretion with respect to both procedure and content of a decision. For many categories of staff the freedom to initiate action is no longer the same. There are indications that middle management in firms may well lose some of its prerogatives when clerical work is partly automated.

Similar implications have not necessarily emerged in all organisations employing computers, but they are widespread. Technological change in organisations is often taken as a pretext to force staff into predetermined work roles. This helped to better obtain their compliance with the rules of the organisation and to secure the conformity of their actions to its goals.

It is still subject to debate whether present developments in computing, made possible by progress in microelectronics, will not completely change the situation. Computer aids at the individual working place could indeed contribute to reversing the tendency to split up work by restoring the continuous processes that have been characteristic of much clerical work before the advent of the computer. Still, the expression "personal computing" often used in this context, may be somewhat misleading as the necessity to interlink work places will no doubt remain and continue to limit the autonomy of workers.

In addition, personal computing in organisations may even further erode personal autonomy by intensifying in-plant control of working behaviour. Decentralised computers can be interlinked so as to make information from many parts of an organisation centrally available. The interaction of people operating electronic devices is often programmed so as to leave a trail behind, without special provisions for erasure. Examples are the automatic registration of telephone calls, the monitoring of set-up times for terminals, etc. Already now the more widespread use of such devices can be tantamount to a complete control of individual working behaviour.

There are several reasons for management making use of such control opportunities. One is better control of individual work performance for the preparation of career decisions. Another is the desire to avoid losses of time spent in activities which management might perceive as useless. Without adequate provisions to prevent such a collection of information about human behaviour and its use or misuse for certain purposes, a closer surveillance affecting the daily conduct of workers may well be the final outcome of a development that originally was justified by pointing, among other things, to enhanced job satisfaction.

ADP in organisations has not yet reached its climax. The example of adverse effects of decentralised computing in organisations indicates that it can be misleading to predict the effects of information technology in the office of the future from the way in which clumsy and expensive central computer systems are still used today.

The following discussion of public administration serves as an example of how a technology like the use of computers has first come to be used in fairly unimaginative ways. This leads to societal consequences which in part may disappear in the longer run. However, they will still be characteristic for some years to come. At least in Western Europe, there is a marked reluctance to make full use of the more recent opportunities for bringing information processing capacity to the individual working place where it could be combined with facilities for data communication as well as for oral and written communication. Obviously, constraints intervene which originate from managerial policies rather than from the available technology.

The case of public administration

Automated data processing (ADP) in public administration did not contribute, as many expected, to bringing about completely new ways of getting things done. Rather, it helped to streamline existing procedures without departing from firmly established bureaucratic practices. The application of ADP still chiefly concerns those branches of public administration which handle routine matters like tax assessment, welfare benefit computation and allocation, etc. Besides, human service organisations, e.g. in the health sector, have computerised a good deal of their auxiliary administrative functions.

Most of the applications of computers in public administration can be reduced to some typical patterns:

(a) *Operative mass systems,* largely performing calculation and logical operations in isolated though often fairly complex routine operations. An example of this is the automated assessment of income tax.

(b) *Integrated systems,* typically support routine operations by providing a common data base for the administrative agencies in charge, or at least providing for some form of data sharing. Compared to type (a), better data storage and retrieval capacity is required. Other agencies may be given access to the data for investigation or for planning or statistical purposes. A good case in point is the Swedish motor vehicle registration system. In addition to the automated registration procedure, it handles taxation, insurance and safety inspection matters, and it even provides the information for manufacturing car registration plates.

(c) *Decision aids,* include information systems of the "data bank" type, storing large amounts of data and generally permitting either their fast retrieval (e.g. in law enforcement or "intelligence" information systems) or their combination for statistical and planning purposes. In both cases, they serve as aids to human decision makers, be it the policeman on his beat or a politician.

Presently, even in countries where ADP already has a deep penetration in public administration, all three types of applications are available only as isolated systems, neither covering the whole body of public administration, nor being interlinked to any greater extent. Pretentious concepts of "Management Information Systems" for the public sector did not bring the results envisioned in the 1960s, although their promise did contribute to the present amount and shape of automation in the public sector.

The societal implications of ADP systems of the types described are already now manifold. Besides their organisational implications, as mentioned before, they affect:

— the *quality of administrative services* as well as the general relationship of public administration with citizens;
— the power balance among public and private organisations (bureaucracies), i.e. the *centralisation* issue;
— the power balance between bureaucracies (both public and private) and the individual, i.e. the *privacy (data protection)* issue.

Before dealing with each of these implications, a general remark is necessary. Information technology, especially computing, should not be viewed as a technology with fixed patterns of use and consequences. Computing is selectively exploited, as one strategy among many, for organising work and information. We are dealing not with necessary implications of computing in general, but with the presently existing uses in a particular field. Available experience tells much more about the strategies and philosophies of individuals and organisations introducing information systems than about general characteristics or inherent properties of the technology involved. In addition, the particular cultural context of countries heavily impinges upon the way information technology is used.

The changing quality of administrative services

In many countries the never-ending complaints about an unresponsive computerised administration convey a largely negative impression. Examples abound of undecipherable computer print-outs, of difficulties of getting computerised records corrected, of outcomes for which nobody seems to be responsible. This should not make us forget, however, that there have been positive effects, too, especially concerning the speed and reliability of administrative work.

It is fairly common for administrative agencies to take the automation of some of their work as a pretext for asking taxpayers or welfare recipients to fill in still longer forms more carefully than before. Such a practice of shifting costs and inconveniences to the client is likely to accentuate existing inequalities in the opportunities of access to administrative services. Automated data processing (ADP) systems frequently demand more skill and higher expense in accessing an administrative service. In addition, motivational barriers become higher as opportunities for direct contact with officials diminish. In many cases, such a reduction of direct contacts has led to a growing unresponsiveness of public administrations to the wants and needs of their clients.

Another negative aspect is the exaggerated reliance of many administrations on internally stored information. With decreasing responsiveness and the diminution of face-to-face contacts between officials and clients, administrations tend to rely more on stored information and on internal information sharing and exchange. Information on individuals is often accumulated without their being aware of it. As the individual has no chance of detecting possible errors and getting them corrected, such information is sometimes quite unreliable.

All this is in sharp contrast to the desire of many countries to use ADP systems as a panacea against maladministration. In some countries in Europe and in the Third World where ADP is less widespread, the view is still held that automated systems might contribute to a more equal and dispassionate treatment of citizens, reduce opportunities for tax evasion and enhance compliance of citizens with administrative law. This view is supported by the observation that centralised systems tend to reduce both factual and legal discretion of officials. There may be countries like Italy where some services were significantly improved in this way. However, similar successes may, in many countries, turn out to be quite short-lived, unless value changes bring about a different climate in the relationship between the state and the individual.

Centralisation of administration decision making

In the debate on the centralising effects of computer use it is commonly held that centralisation implies power shifts in favour of the top of an organisation or of a central organisation, whereas decentralisation would imply the reverse.

Power shifts, be they intended or not, often have to do with better information available at the centre of an organisation or a network of organisations. Central data storage, or at least the possibility of drawing the decentrally stored data together instantaneously, is likely to strengthen centralised control. Yet the related discussion is still lacking in conceptual clarity. Too many factors independent of information technology are in fact involved. Little effort has been made to identify the relative contribution of information technology to social change and power shifts. It seems as if the easy-at-hand and seemingly convincing formula "information is power" detracts from a closer look at the nature of the problem.

While there is some evidence of power shifts *within* organisations, the discussion related to the power balance between government and administrative organisations still remains an extremely hypothetical one. One field of concern has been the relationship of the executive branch of government to parliaments where a weakening of parliamentary control due to the emergence of administrative automated data processing (ADP) systems has been feared.

To address power shifts among and within organisations in terms of centralisation and decentralisation can be misleading unless it is borne in mind that "centralisation" can have quite different meanings. The centralisation of routine tasks merely for reasons of expediency is something else than a centralisation consisting in a reallocation of decision-making power which implies a loss of autonomy, discretion and power at lower levels. Still another form of centralisation would be the merely physical centralisation of data storage, of computing operations, or of systems analysis and programming.

The belief that computer use would have centralising effects was long motivated by economies of scale made possible by centralised computing.[2] The efficiency of early computers increased rapidly with increases in their size. This seemed to call for centrally located computers to be shared by a variety of administrative agencies rather than each agency using its own smaller computer. A second characteristic

---

[2] Herbert Simon, The consequences of computers for centralization and decentralization, in: Dertouzos and Moses (eds.), *The Computer Age, a Twenty-year View,* pp. 212-228 Cambridge (Mass.), MIT Press, 1979.

of early computer systems was that they could be used only in physical proximity, which would centralise that use.

With time sharing and remote access to computers by communication lines, many users in different locations can share the same computer. The decision-making process can thus be almost completely detached geographically from the place where the computing is done.

These elements are of importance with regard to the ongoing debate about reversing the trend of increasing centralisation which so far seemed inescapable when computers were used to any greater extent. If computing has so far strengthened centralised decision making, this may no longer be the case. It would be misleading, however, to suppose that the new technology has a bias in favour of greater decentralisation. Although technical feasibility and economic opportunities play an important role in shifting the locus of decision making to higher or lower levels, a decentrally operating technological device is not equivalent to decentral decision making in the sense identified above. Decentralised computing in the current sense obviously will not automatically lead to more autonomy, power and discretion at lower levels.

Instead, to an increasing extent, central control over information and freedom of action of decentralised bodies will coexist. Already in the past, in many cases, computers were not installed for increased efficiency alone. Rather they seemed to hold the promise of better control over important operations. With the support of automated systems, the upper levels of authority are in a better position to control the work being done in their own unit or in decentralised units under their supervision. Moves in the sense of conferring more power to decentralised units can be accompanied by better informing the centre about their operations. Thus if the new technology has a decentralising effect, this could as well enhance central power instead of weakening it.

Privacy and Autonomy of the Powerless Individual

Whereas the centralisation of information management and of information resources has to do with power shifts between and within organisations, the issue of data protection, or protection of personal privacy against intrusion by information technology, is related to the power balance between organisations and the individual. This has not always been very clear. The real issue at stake is not personal privacy, which is an ill-defined concept, greatly varying according to the cultural context. It is power gains of bureaucracies, both private and public, at the expense of individuals and of the non-organised sectors of society, by means of the gathering of information through direct observation and by means of intensive record-keeping.

Origins of concern

Concerns about the invasion of privacy by computes were first raised in the 1960s. It was feared that organisations using automated record-keeping systems would collect more sensitive data about individuals than before, and share these data more easily with each other. It was anticipated that this would lead to serious threats to personal privacy. A proposal for a "National Data Centre" put forward in the United States sparked off a whole series of more or less speculative studies. These dealt with the dangers looming when organisations would be provided with a longer and more accurate memory, and be able to draw on pools or data which were previously not accessible to them.

Yet the forecast was uncertain. Today there are still no massive centralised data banks which could pour out detailed life histories of very large numbers of people at the touch of a button. However, many specialised systems of this kind do exist in many countries. Examples are:

— personnel information systems, especially in large firms or in government, e.g. in the British Civil Service Department;
— health information systems, e.g. in hospitals;
— social security information systems;
— criminal information systems, containing, for example, criminal histories, data on drug addicts, wanted persons, etc.

Such systems sometimes do not store all information in a machine-readable way. Especially in the police sector, they often indicate where a certain information is to be found in written records kept elsewhere.

Nevertheless, systems like these have no doubt a certain potential for adverse use concerning the individuals whose data are stored in them: whether this adverse usage is deliberate or whether it is by inadvertent leakage. The individual is likely to be confronted with better informed organisations and professionals.

A frightening perspective is the possible coalescence and interlinkage of such systems. Until quite recently, in many countries one could be fairly sure that individuals were known by administrative agencies only in determined roles, e.g. as taxpayers, patients, pupils, employees. Without adequate provisions to the contrary, data from different sources could be drawn together more easily in automated information systems. Like a puzzle, they would constitute a rather complete but probably also distorted image of the personality of citizens. Such a danger is likely to grow with a better command of computing, especially with regard to the linking of computers in communications networks.

Privacy legislation

Again, the case is one of confronting fears with reality and with the presently observable trends in that reality. The last decade has provided pervasive evidence that computers clearly are not, or at least not yet, the powerful governing machines they promised to be. At the same time, it became obvious that some of the fears about the invasion of human privacy by computers were misplaced. Measures to protect the individual against computer domination were often designed very hastily. There is no doubt that they contributed to increasing sensibility with regard to the problem. But despite the will of its enactors, the existing privacy legislation tends more to calm down public distrust roused by plans for "government by computers" than to provide for effective protection. Regulating computer use only, as is the case in most national data protection laws, may well fall short of solving problems which have been exacerbated by computer use but rarely originated by it.

Many countries have enacted legislation to prevent misuse, or unfair use, of computer-stored data. Since 1970, such legislation has covered several West European countries, as well as the United States. It generally tends to take the position that there would be no reason for the public to worry as long as record-keeping systems allow the persons concerned (the "data subjects") to check the accuracy of their records and to be kept informed about the records' use, at least in the majority of cases of that use.

Carefully designed procedural safeguards are characteristic of the conception and structure of most privacy legislation. Only occasionally do these laws spell out in detail which data may be kept and processed in determined ways for determined purposes, thereby posing limits on the kinds of record-keeping systems that should or should not be built. Instead, most of the legislation is aimed at setting out rules on the supervision of data processing installations. The laws are furthermore very explicit about questions of access and correction of data by the persons concerned. Usually, they cover only data that are processed automatically, but some go further and include conventionally stored and manually processed data as well.

It may be doubted whether in this way legislation can neutralise the challenge to privacy and individual autonomy. With regard to the situation in the United States it has been argued that, by defining a model of protection based solely on individual procedural rights, such a policy has promoted rather than forestalled the trend to mass surveillance in society.[3]

First experiences in implementing privacy legislation in several countries soon made clear that there was often overlap or even contradiction with other parts of the national legal system, where it concerned the handling of information, be it in an automated way or not. The old debate on the amount of secrecy which should be afforded to protect the vital interests of a country and of its power structure has been reinvigorated by privacy legislation. In some cases, the rules enacted directly counteracted the principle of freedom of information. In many respects, they seemed to further weaken the opportunities and rights of open access to government records which in many countries are believed to be an essential prerequisite for a free and open democratic process. Also, privacy legislation has served in many cases

---

[3] James B. Rule, in Lance J. Hoffman (ed.), *Computers and Privacy in the Next Decade,* New York, Academic Press, 1980.

to curtail the information base of critical social research, especially on subjects of public policy.

Yet in several countries, including the United Kingdom, France and the United States the balance is now swinging to the other side. After some overreactions in reinforcing existing secrecy provisions which were presumedly required for protecting individual privacy, there is a marked tendency towards greater openness in handling information, at least as far as the public sector of these countries is concerned.

One of the effects of the privacy debate is a greater consciousness in handling person-related information. This debate has raised the information management practices of corporations and of public agencies to a politically visible level. The effects are not only a heightened interest in the individual's right over his personal data but also a strengthening of the freedom of access to government records by the general public.

Relatively vague provisions about which information would have to be made accessible to whom are gradually being replaced or supplemented by more elaborated provisions concerning the use of health data, criminal records, social security information, etc. Concerns about the invasion of individual privacy by information technology are thus gradually transformed into concerns about important sectors of social life where bureaucracies gather information about individuals for purposes of social welfare and of social control of individual behaviour.

Privacy and the power of bureaucracies

Even though there are similar debates about the policies of information provision to bureaucracies, many knowledgeable specialists regard the problems generated by information technology as being settled with the enforcement of appropriate legislation. Yet the problem posed by the better availability of information resources for organisation decision making has many facets. Privacy legislation tends to comfort itself with the provision of rights for citizens concerned, to check the correctness of files and ensure fairness of information collection and use. It cannot answer the question of whether better availability of such information resources will not lead to power shifts in favour of large bureaucracies at the expense of the individual. If such were the case, individual rights would clearly be insufficient for redressing the balance. Instead, new forms of control of such large bureaucracies would have to be found.

In short, the problem has been too hastily stated in purely individualistic terms in order to make it manageable within national legal systems. Safeguards were devised before the real dangers at stake were sufficiently identified. Little effort was made to assess the relative impact of automated data processing (ADP) on changes in the patterns of information use and information flow. It was not made clear to what extent the breakdown of formerly "natural" barriers to information flow was due to better technical means to overcome such barriers, or whether its main stimulus was a marked desire to obtain more information for making better decisions and for intensifying centralised social control.

The effects of the changing information balance are manifold. They encompass far more than encroachments upon personal privacy. Civil rights, freedom of expression, the possibility of taking part in political discussions free from fear of negative reactions, are some of the fields where a changed information balance may produce adverse impacts.

A deeper analysis of the problem should start with the fact that any social interaction implies the transmission of information about individual behaviour. This applies regardless of whether such an interaction is direct, spontaneous and informal, or channelled by rules and mediated by bureaucracies. Behaviour which is considered as deviant behaviour by some participants in the interaction, especially the more powerful, will provoke reactions of social control, negative sanctions.

Information technologies enter the stage when they contribute either to enhancing the direct visibility of individual behaviour or to facilitating access to stored records about such behaviour. Street surveillance cameras and Orwell's *1984* monitors may serve as an example of the former, whereas access to computerised personal health data by the police illustrates the latter.

Surveillance as the starting point for social control will in both cases be intensified. Apart from this, another effect of information technology is easier communication about deviant behaviour, especially through automated police information systems. All these effects can make social control easier and tighter and provoke changes in the patterns of sanctions of deviant behaviour.

It is astonishing that so few efforts have been made clearly to identify the benefits and dangers of changed patterns of information flow, of information control and of surveillance. On a strictly individual level, some of the potential dangers are quite well known, including discrimination by employers or credit firms as well as the impossibility for an individual to make a fresh start. Of equal importance, however, are societal and political consequences. Changes in law enforcement patterns may be welcomed as far as they contribute to more equal treatment of citizens in their relations with public administration or to filling gaps where in our present Western societies social control is dramatically lacking. Yet there is the risk of a disproportionate increase of social control in circumstances where deviant behaviour is of minor social importance but easy to observe and to register.

So far, the availability of police information systems primarily resulted in greater efficiency in tracking petty offenses, helping patrolmen to find stolen cars or unpaid parking tickets. Quite often this has resulted in displacing time and attention from other important police tasks. Rather than more intensive social control we might therefore expect a new distribution of control activities, especially more centralised social control. In many cases such a shift in the locus of social control from local institutions and informal groups to central agencies seems to be independent of technology as such. But in other cases it would appear that information technology creates an opportunity for centralisation of social control, especially by making human behaviour more visible and by multiplying accounts of such behaviour that may not easily be "forgotten" by the controlling institution.

In the long run, information technology will probably contribute to the intensification of social control, not only its centralisation. The question is open, however. Social forces might arise which would fiercely resist this. It may even be that there are limits of a more fundamental kind to the intensification of social control. Removing the obstacles to full visibility of human behaviour may ultimately demand a price which no society would pay. So far, perfect communication, the full visibility of behaviour, has been characteristic of many

Utopian models of society. It is also the ideal of prison architecture, but it never worked in real life situations where some opportunity to withdraw always had to be provided.

The Marriage of Computers and Communications

Most of the implications mentioned so far are due to computing, to automated data processing (ADP). Other information technologies, especially telecommunications, only occasionally play a role. The range of societal implications of information technology is considerably enlarged, however, as soon as the coming together of elements of computer and communications technology is taken into account. The existing networks for written, oral and visual communication are as yet one of the outstanding application fields of microelectronics. They are supplemented by data communication networks. Geographically dispersed computer systems are linked to each other by data transmission facilities and are made accessible to users from remote terminals.

This development opens up the perspective of computers and new forms of communication infiltrating everyday life, as did the telephone and, in some countries, the typewriter. Not unlike the supply of gas and water, "information utilities" would eventually bring a wide variety of information access opportunities, as well as services, to the home. After the rationalisation of production and of clerical work, such a development could constitute a new wave of rationalisation affecting everyday life.

Information utilities are still at a very early stage of development. In some countries pilot installations of two-way broadband cable communications systems can be found. They provide an infrastructure for a wide variety of communication processes and information services. Comparatively well advanced are systems for information access via telephone, using TV sets as display devices, like the British PRESTEL systems. Compared to the few concrete realisations there is an enormous amount of speculation about the potential benefits and harms of information utilities.

We will refrain from repeating here much of the current speculation. Serious doubts must be expressed as to whether information networks as envisioned in scenarios like that of a "Wired Society"[4] will actually work as supposed. We do not know, for example, whether one day there will be automated voting and polling systems, with the impacts on the political process described below. We do not know very much either about the new cottage industries made possible by communication facilities, linking the home to offices and colleagues.

Scientific and technical information systems

One field where computer/communications networks are already being used to a greater extent is the provision of scientific and technical information accessible from large computerised data bases. Information systems for scientific and technical information accessible from remote data terminals are in use in several countries. Although there was no breakthrough as expected some ten years ago, we will probably soon get used to scientists working at a remote access console searching for data, references and the full text of articles. The exact level of eventual usage of these systems is still unknown. Part of their potential benefit might perhaps be provided by home computers coupled with video storage media, which would confine their use to timely information and interactive services.

In any case, contrary to some speculations, scientists will continue to read books. This depreciates much of the speculation about huge data banks constituting monopolies of knowledge. The instant rewriting of history following the directives of the day will probably be no easier with such systems than it is today. More generally, anticipated consequences of the "uniqueness" of automated information systems will no doubt be tempered by the continuing use of alternative media.

An important problem of scientific and technical information systems is connected to the opportunities of accessing them. Access may be limited by imposing legal and financial restrictions on potential users. Preferential access might be granted to privileged users like government agencies and large corporations, which no doubt would

---

[4] James Martin, *The Wired Society,* Englewood Cliffs, Prentice Hall, 1978.

give them comparative advantages. University research is already lagging behind corporate and government endeavours in many fields. Such a situation could deteriorate even further. Similar questions arise with respect to the access of Third World countries to information systems still located almost exclusively in the United States and Europe.

To redress the balance, however, legal provision of access rights, and economic regulation to ensure access for all potential users are not sufficient. Often enough, available information systems are biased in that they contain information specially adapted to the needs of specific user groups only. In addition, skills to make use of the systems are unevenly distributed. Unless the educational system adequately responds to this challenge, equal access to information is certainly not tantamount to equal opportunities for enhanced knowledge. Instead, new access opportunities might only result in further information overload for the individual, even the highly educated one.

Coping with information overload is obviously impaired by the lack of attention which the potential recipient of information can devote to it. Although information systems could in principle also act in the sense of providing instruments for better coping with information overload, criteria for selecting information will continue to be formed rather by personal communication.

To sum up, there are a number of indications that the benefits of scientific and technical information systems will accrue selectively, with large organisations and the professions reaping higher benefits than the general public.

Information utilities

The emergence of information systems not restricted to scientific and technical information but geared to what their proponents perceive to be the needs of a general public might well make no difference in that respect. If they are to work as envisaged, they will probably deepen the existing cleavages between the educated and the less educated. First developments coming close to the information utility concept are the national videotex systems presently developed in several European countries as well as in Canada, the best known being the British PRESTEL system. They are designed to provide access to information and services in many fields. Whereas the technical end-devices (TV set, telephone plus a modem) are commercially accessible to many inhabitants of the Western world, it remains to be seen whether these systems will become as widespread as the TV and the telephone. There seems to be a great potential for social stratification, depending as much on education and skills to make use of the system as on the ability to pay for the different services. Only a fraction of the population will benefit; those people not able to deal with the new technology will fall still further behind.

Information systems for general use pose the problem of access to information in a much more intricate way than scientific and technical information systems. If such systems were to be relied upon to a considerable extent, their potential to shape the information environment of people would at least equal that of the mass media. Legal rights to information access, as well as existing forms of regulation of the mass media, would not suffice to redress the situation. Rather, educational efforts aimed at developing the ability to make use of information in a meaningful way would be required in order to counteract the potential for political manipulation.

Another implication commonly anticipated is the isolation of individuals. Implicit in many of the designs of information utilities is an atomistic view of the individual sitting at his home information terminal. It might then be feared that talking to machines instead of other human beings might further isolate him.

There is, however, not enough conclusive evidence in support of such an effect. On the contrary, the absence of face-to-face contact in many settings probably does not lead automatically to more isolation. Research on the effects of telephone conversations indicates that these are not a weak version of face-to-face discussions. Contrary to expectations, the telephone released unsuspected powers of speech from the constraints imposed by face-to-face conversations.[5]

[5] Fred Emery, Communication for a sustainable society, in *Telecommunications Policy*, vol. 4 (1980), p. 306.

A more important source of isolation of the individual is no doubt the growing impersonality of the provision of services. Information technology can contribute to this in many respects, the best known being probably the changes in banking, with automatic tellers replacing clerks.

Politics in the "information society"

Among the speculations about possible societal benefits of information utilities there are many plans for their incorporation into political life. If every home were equipped with a communications terminal connected to a nationwide network, there would be opportunities for improving the communication flow between ordinary citizens and political decision makers. This vision of the future forms the background of plans for national plebiscites at regular intervals or other citizen response systems where the government would elicit responses on particular issues.

The different forms of "electronic voting" have several implications generally left unclear by their proponents. They would strengthen vertical communication linkages between citizens and decision makers, instead of enhancing existing horizontal communications structures and the ability of societal sub-groups to express their opinions.[6] Underlying this is the atomistic assumption of isolated citizens confronting and deciding the issues in their respective homes.

Active political participation in "instant referenda" would probably benefit the existing power structure more than the citizens. The relevant questions to be answered are defined and the conditions of feedback dictated by the power structure. Rather than providing true participation opportunities, this would at best reduce the sense of alienation of citizens, as is the case with presently available forms of telephone feedback to television stations. There would be a temptation to elicit immediate reactions from the constituency which would reduce rather than increase the amount of political discussion. Oppor-

[6] Kenneth Laudon, Information technology and participation in the political process, in A. Mowshowitz (ed.), *Computers and Human Choice*, vol. 2, pp. 167-191, Amsterdam, North-Holland.

tunities for legislative innovation might thereby be jeopardised. Moreover, extended forms of electronic voting would provide legislators and governments with information on public attitudes, increasingly enabling them to predict public wishes and reactions, and to anticipate adverse reactions. At the same time political apathy could augment if everyone had to fear that his political opinions were not only listened to, but automatically registered and stored.

Even though electronic voting in the strict sense may not be implemented in the near future, opinion polling by "sweeping" communication networks, as well as the registration and aggregation of consumer choices in remote ordering systems may carry similar implications. Social and market research would be particularly well served by that kind of data, with social scientists more or less directly serving government and corporate interests, and becoming the unwitting promotors of tighter behavioural control and of the existing distribution of power. Certainly, the need for the implementation of privacy safeguards has been loudly voiced. Less well acknowledged, but perhaps of even greater relevance to society is the contribution made by better information resources to the stabilisation of the governance of a polity that is bound to become increasingly predictable in its reactions.

Greater political stability is of course desirable where weak governments are shaken by coups d'état or terrorist attacks, or cannot prevail against Mafia-type semi-clandestine organisations. But the sort of stability eventually attained with the intensive use of sophisticated information technology for political purposes may be such that peaceful political change and societal innovation would be in serious danger.

Visions like the "electronic plebiscite" have a legitimate base. The democratic process poses organisational problems, and the application of information technology can have organisational implications which could make these problems easier to solve. There are two classical answers to the organisational problem of democracy: representative and council democracy. Whereas the benefits of increasing information handling capabilities to the representational process remain doubtful, experiments with forms of council democracy like the "Electronic Town Hall" (Etzioni) could show ways of enhancing participation in political affairs. Yet it would be premature to expect much progress. Alternative forms of the democratic process involving better citizen participation can be conceived of even without technological aids. It is hard to see how information technology could change the existing structures and institutions which have prevented the realisation of alternative forms of democracy so far.

The transborder flow of data

Computer/communication networks are increasingly crossing international borders. Hence the transborder flow of data, although nothing new in itself, is significantly increasing. It makes little difference to the fire chief of a Swedish city whether his data on buildings, access ways etc. are stored in a computer located in the United States. He may even be completely unaware of this fact.

Uses of international computer/communication networks serve purposes of

— remote data processing (the sharing of automated data processing resources);
— corporate management (the co-ordination of production, distribution as well as financial and personnel management within transnational corporations);
— transactions (banking, airline seat reservations, etc.).

There is a growing amount of research and also of speculation on the potential effects of such networks. Quite often, this is lumped together with guesses about the implications of communications satellites, namely the exposure of a country to remote sensing and to foreign mass media and propaganda.

Transborder data flow is to be welcomed insofar as it allows for cultural exchange on an equal footing. However, in many countries (especially in France and in South America) it is felt that national sovereignty is affected if the exchange is one-sided. The operations of transnational corporations would be less controllable. Data would be exported to be processed and stored elsewhere. The importation of scientific and technical information kept in data-banks of the English-speaking countries of the First World would endanger the cultural independence and uniqueness of the receiving countries, as would their exposure to media contents produced abroad.

Among the threats to national sovereignty, understood as a nation's ability to control its own destiny, there is a great vulnerability of national economies and political systems. There might be threats to cut information supply, or computing power, as a means of international politics. This implies the danger of losing access to vital information affecting economic and social development. Such a dependence on foreign resources may constitute a serious threat in times of conflict, perhaps weakening the political position of a country. It cannot protect itself against damage, destruction, or misuse of information stored abroad. Another alleged aspect is the ability of secret services of the country where a data base is located to draw inferences from secret material.

Of at least equal relevance is the opportunity to circumvent national information legislation and regulations. Transborder data flows which are difficult to control entail a decline of opportunities for enforcing national information legislation. Restrictive privacy legislation could thus be evaded by transferring data to "data havens" abroad. Likewise, copyright laws as well as customs regulations could be evaded when information is transmitted (exported or imported) electronically instead of being transmitted on a material support (books, reels of tape, etc.). Although, in the case of privacy legislation, important efforts are being made by international organisations to develop an international legal framework, the problems arising from the desire to channel and monitor more and more volatile information are by no means well understood, let alone resolved.

Threats to the cultural sovereignty of countries are more likely to arise from satellite television than from data communication. The exposure to foreign-produced mass media contents constitutes a threat to the cultural identity of countries, especially in the Third World. However, the cultural identity of a country could also be affected by reliance on foreign data banks. A French journalist searching the *New York Times'* news data bank for factors of French political history will receive a picture biased by the views of his American colleagues.

The way this information is structured influences the conclusions drawn from it.

Information Technology Impact: Myths and Reality

The wide range of societal implications of information technology discussed in this chapter is by no means exhaustive. Many social problems and phenomena could be discussed as well. Yet the further adding of problem areas will bring us to a point where unfounded speculation would take the lead over what can be deduced from available experience. Much of what is expounded in the rapidly swelling literature on the societal implications of information technology is necessarily based only upon speculative thought, as many of the new technologies have not yet been built and located in their social settings.

Speculation, of course, cannot be avoided. Nor would it be desirable to do so. The long-term effects of information technology on society are fairly unpredictable. There is no reason to conclude that there will be no such effects, simply because there is as yet little evidence. Computing and the merger of computers and communications are relatively recent technologies, and their real and lasting impacts are emerging only very slowly. The serious and intractable problems of information technology might even not appear for decades.

On the other hand, our analysis is in constant danger of being misled by biased perceptions of the opportunities and risks of information technology use.

It is a recurrent phenomenon that new technologies are welcomed as powerful agents that might bring about thorough changes in the structure of society and in the living conditions of mankind. As so many social problems apparently resist conscious efforts to tackle them, many hopes for a better world concentrate on technologies. It is not astonishing, therefore, that many emerging technologies are welcomed as panaceas for all kinds of societal diseases. But in the case of many new technologies such periods of hope were followed by bitter disillusion.

The computer, as the core element of microelectronics-based information technology, has already given rise to phases of hope and disillusion. The only difference, compared to other previously developed technologies, is in the emphasis of the Utopian thinking which it has fostered. It is negative rather than positive.

An analytical effort is required in order to escape from the dilemma of pessimistic fear of negative impacts and optimistic belief in the convergence of technological and societal progress. Such an effort still seems to be absent in much of what is being written on computers invading private lives, on one side, or on an electronic paradise free of toil, on the other. Our reflection capacity is literally overwhelmed by our anxieties and hopes.

The computer as a projective screen

Like other technologies before, information technology and especially the computer act as a projective screen for other social and political concerns.[7] A discourse about computers can carry feelings about political and personal issues, e.g. anxieties about not feeling safe in a society that is perceived as being too complex. As bureaucratic power almost by definition is impersonal in many settings, no agents responsible for the repression of human aspirations can be identified. Anonymous structures limiting human freedom are more and more replacing the direct wielding of power.

Feelings of being watched, even persecuted, therefore focus on the computer which more and more comes to stand as a symbol of bureaucratic power. Because of its easily identifiable role in corporate and government bureaucracies, the computer attracts the anger and frustration which are endemic to life in a society dominated by bureaucracies. Our deep-rooted wish to identify causal relationships makes us blame the computer for the alienation of individuals, their being left alone by an unresponsive, impersonal society.

In this way, the computer can act as a smokescreen as well. Computers can serve as an explanation for social phenomena that might otherwise raise disturbing questions. Attention is being drawn away

---

[7] Sherry Turkle, Computer as Rorschach, in *Society,* vol. 17 (1980), No. 2, pp. 15-24.

from the underlying issues and onto the debate "for and against computers". This may well detract from closer thinking about these underlying issues.

Another reason for overestimating the role of information technology in social life can be found in its challenge to human self-understanding and self-esteem. The anxiety over the diminishing role of human beings concentrates on computers and robots. The perspective of artificial intelligence forces us to re-examine our belief in the uniqueness of human attributes.[8] Intelligence is exhibited in the behaviour of computer programmes which prove mathematical theorems or recognise patterns. The set of attributes defining man's unique endowment grows smaller and smaller. Contemporary fiction best expresses the anxieties connected to this shrinkage when it foretells the creation of intelligent computers or robots which will rival and perhaps replace man as the dominant species on earth.

The demonstrable accomplishments of information technology are only partly responsible for the challenge to human uniqueness. Information technology also influences the metaphors and concepts we use to understand ourselves. Our willingness to accept comparison on the computer's terms is deeply rooted in Western rationalist tradition. The tendency to think about ourselves as if we were mechanical devices not only pervades popular thinking, it also influences the behavioural sciences. Information processing concepts have been applied to the study of perception and cognition, as well as to the study of societies as information processing systems.

Technology use reinforces social trends

In any case, the tendency to blame the technology for social problems is not only due to our desire to single out a scapegoat every time we do not fully understand social phenomena. There is also a marked tendency of computers to increase the urgency of many social problems.

---

[8] Abbe Mowshowitz, *Inside Information—Computers in Fiction,* Reading (Mass.), Addison-Wesley, 1977.

Paradoxically, they often do so in lending strength to socially accepted ideals. The desire for homogeneous conditions of life, for equality and more orderly societal arrangements is a powerful stimulus towards an ever more perfect administration. Information technology, in many respects, is of great help in implementing such a better administration. It can help in making dreams of a more perfect world come true—to an extent that the negative sides of such a perfect world become really visible for the first time.

A good case in point is law enforcement. Better tracking of criminals seems to be a goal universally agreed upon. And yet, we do not know to what extent societies need leeways for "deviant" behaviour in order to stay open for evolution. We do not know either whether some degree of non-prosecution is not required for ensuring compliance with social norms. It it were generally known to what extent many norms are continually violated, then it might well be that compliance with them would be greatly reduced.

One important question now is whether all kinds of social trends can be corroborated by information technology in the same way, or whether this technology acts selectively. Another way to put this question is to ask whether all social forces can successfully employ information technologies for furthering their own aims, or whether some interests reap higher benefits from its use than others.

Information technology enlarges the weaponry of social forces

There seems to be a ready answer to this question. It takes its start from the fact that technology in general takes its shape and finds its application areas through the action of social forces, their aims, conflicts, and ideologies. There is an ongoing struggle in society for power and prestige. Information technology enlarges the weaponry of the conflicting social forces. Past struggles and investments which have already been made set the stage on which new information technologies appear in society. In the legal and institutional framework thereby constituted, the unequal distribution of economic opportunities and (other) power positions inevitably leads to new technologies being taken up first by those who can afford them. Their use and indeed their very development would in this way be biased so as to serve primarily the interests and purposes of the mighty and the wealthy. The "Matthew effect" ("For whosoever hath, to him shall be given...") appears as the most outstanding implication of new information technology, making the already powerful still more powerful.

Such a view on the question of the societal implications of information technology is becoming increasingly popular. It corresponds to the impression many people in the industrialised countries get from observing the changes taking place in their immediate environment. To them, private and public bureaucracies seem to be farther away—less accessible than before—when administrative processes are handled by computers.

In addition, such a view has the important side-effect of keeping automated data processing (ADP) neutral and immaculate. The engineer may continue to present his favourite toy as a tool to a society, the actual working of which he ignores. At the same time, he can react against presenting computers as the scapegoat for outcomes which are the results of the policies of those introducing and using computers.

Lastly, this view has the further advantage of making the problem manageable within the dominant frames of reference of the social sciences. Society is generally conceived as an arena where struggles take place leading to a continuing reallocation of both material and power resources.

Yet there is an important flaw in this argument. The continuing redistribution of resources may well detract attention from an evolution of society which is slowly taking shape through these redistributions, i.e. the accumulation of power in the organised structures of society. Conceiving of such power shifts in a frame of reference which assumes the steady repetition of political struggles may hide the possibility that these might end one day or at least lose their significance.

A more orderly society may indeed be the long-range outcome of changes which are only beginning to take shape. The great question behind it is whether this order will be limited to the realm of necessity or else impinge on man's creativity, his judgement and his freedom, leading ultimately to a completely regulated, bureaucratic society.

Inherent properties of information technology

From these critical observations on one way of conceiving the societal implications of information technology we shall now turn to a competing tradition of thought. In such a tradition, it is not power that is thought of as the primary force of social change. Rather technology itself plays the dominant role. A technology may indeed have inherent properties which make it useful for attaining certain purposes.

In order to assess these inherent properties, we have to come back to the distinction between automated data processing (ADP) and communications technology. Mass communication technologies and other communication technologies permit human exchange in spoken or written form or by images. Their important feature is their character as media. Hence many of their possible societal effects can be associated with the isolation of individuals, the loss of face-to-face contact.

Compared to this, ADP clearly has an additional feature. It serves for processing (storing, combining, transforming and forwarding) highly *formalised* information. This implies of necessity that this information is removed from its pragmatic context while being processed. When the output is then fed back into a real life situation, it will only be of use if such information has a clear and unequivocal meaning. Compared to human information handling, there is an enormous loss of flexibility which makes ADP ill suited for all social situations that cannot or must not be formalised. It inevitably distorts the physical reality, transforming it into a rational world, removed from the real world which is only partly rational.

Computing has an important bias. It handles formal symbols only, and in a perfectly rational way. It strengthens rational thinking at the expense of non-rational, creative thinking. Human thinking can be divided into two different modes. One is the rational, analytic mode, associated with calculation, reading and writing. The other is the intuitive, holistic mode, associated among others with pattern recognition and musical ability. It becomes apparent that the computer unwittingly imbalances our mental processes by strengthening one of these two complementary modes of thinking. Its application in human affairs introduces a mechanistic rationality into social relations of a predominantly informal and non-rational nature.

Information Technology and Bureaucracy

Looking at the inherent properties of information technology explains more of its success and its societal implications than conceiving of it only as an instrument available to social forces. The extent to which formalised structures dominate a society is not unaffected by the technologies of which this society avails itself. Perhaps the most general feature of the beginning "informatisation" of society is the growing significance of formal elements in the social fabric. Information technology is simultaneously the immediate cause and the vehicle of their introduction.

Bureaucratisation and informatisation of society:
two interrelated processes?

The triumph of the computer finds a striking parallel in the growing role of formal organisations, of bureaucracies, in society. The bureaucratisation of society is the point from which a macrosociological analysis of the societal implications of automated data processing (ADP) could take its start. Bureaucratisation as understood here is more than just the existence of large organisations with hierarchical structures and other characteristics of bureaucracy. It means that an increasing part of social relations is constituted by human interaction via bureaucracies, by role-taking within or in relation to bureaucracies. As bureaucracies assume more and more functions of social service and control, they can be seen as nodes in the social fabric and as a major factor of social integration. In a society dominated by bureaucracies, many human relations lose their direct character. They are mediated by organisations which more or less have the power to regulate human behaviour.

Bureaucracy, in this view, is the ultimate force holding together the parts of an atomistic society which increasingly depends on it for its vital functions. Bureaucratisation, as understood here, is a much more recent development than the birth of bureaucracies as purposeful and formal organisations. The presence of public bureaucracies in everyday life, their pervading of all politically relevant social structures is of very recent origin. Even in the model of the bureaucratic State, nineteenth-century France, the interventions of public administration still had an exceptional character, from the individual's point of view.

Informatisation and bureaucratisation of society are congenial processes. Both are rooted in the apparent superiority of formalised structures for achieving particular ends. This obvious parallelism of informatisation and bureaucratisation strongly indicates that the lead of organisations over the unorganised segments of society in making use of information technology is not exclusively due to economic or political reasons. By its ability to process formalised information, ADP can be particularly helpful to bureaucracies which tend to formalise both their internal structures and the perception of their environment. ADP contributes to enhancing bureaucratic power at the expense of the unorganised segments of society, as it confers increased efficiency to the bureaucratic mode of producing goods, administering, rendering services and solving problems.

ADP is important to bureaucracies because it is the embodiment of rational administrative behaviour directed at the efficient attainment of organisational goals. It is a powerful aid at the service of the bureaucratic mode of coping with social problems of all kinds. Computers tend to reinforce the secular process of bureaucratisation of the society as they enable bureaucratic organisations to respond to rising demands and increased societal complexity. In doing so, they contribute to making individuals, as well as society at large, still more dependent on the bureaucratic organisations that structure social intercourse and cater for many human needs.

Information technology makes society more vulnerable

The dependence of society on large bureaucracies for some of the most vital functions and services also implies an increased vulnerability. There might be events where some or all of the bureaucracies would cease to work. The perspective of a possible breakdown of centralised computer systems, on which bureaucracies have come to rely, significantly adds to the vulnerability of a society depending in many respects on centralised bureaucracies. Such a vulnerability is greatly enhanced when society depends on large integrated networks for many of its needs. The consequences of the breakdown of networks are of a more serious nature than those of the breakdown of decentralised services with a low degree of interdependence.

An example where information technology brings about a quantum jump in the networked nature of services is financial transactions. Electronic funds transfer systems involving automated bank account operations are implemented to upgrade the existing giro systems, making their use increasingly popular. Later, linkages to automated points of sale would permit shopping without cash, cheques or credit cards.

Systems like those developed for electronic funds transfer present dangers in several respects.[9] They make possible new forms of criminality and present new targets for terrorist attacks. The destruction of data files from remote terminals is one example of deliberate obstruction of such systems, the breakdown of which could have serious economic or social consequences.

A major source of technological vulnerability lies in the sheer size and complexity of many automated information systems. Many systems with a high degree of interdependence are built without previously answering the question of how much protection from a major fraud or an armed attack is required, and how much redundancy for backup is necessary for recovery of part or the whole of a network.

The incomprehensibility of large systems sheds doubt on the belief that computing is actually under human control. In 1980, two systems failures leading to military alarm have again posed the question to what extent is it reasonable and safe to rely on highly automated and complex systems for the management and monitoring of critical tasks. In some fields including decision-making processes deeply affecting individual life situations, a moratorium on building large, incomprehensible and dependency-creating systems would be the appropriate reaction to the present situation.

Vital decisions affecting the survival or life situation of human beings should not be left to machines alone. They continue to require human intervention.

---

[9] Donn Parker, Vulnerabilities of EFT systems to intentionally caused losses, in Kent W. Colton and Kenneth Kraemer (eds.): *Computers and Banking,* pp. 91-102, New York, Plenum Press, 1980.

Towards a perfect administration?

Presenting information technology as a factor contributing to increased social vulnerability, however, is only one side of the problem. There are many other technologies like nuclear power or genetic engineering that create manifold dangers and thus the need for adequate measures against their misuse. Information technology is unique in the sense that it can also be used for combating social vulnerability caused by other technologies.

An important case in point is access control and surveillance of nuclear power plants including, for example, automated voice and fingerprint recognition. The belief is widespread that nuclear power plants can be better safeguarded when as many measures as possible do not rely on human intervention.

Such a belief, however, is extremely dangerous in itself. It could lead to assuming that installations can be actually safeguarded which, according to our present experience and knowledge, are virtually impossible to safeguard. It could lead to grossly underestimating the risks contained in the "plutonium economy", created by the "fast breeder" nuclear reactor line. As a few kilograms of plutonium would be sufficient to build an atomic bomb for terrorist use, there must be no exception to the keeping track of the increasing amounts of plutonium generated in these reactors. The requisite surveillance system is unprecedented. As any social instability could imply the physical extinction of whole societies, a stable government becomes an imperative, and this for time spans that are beyond any common political imagination. The use of information technology for purposes of surveillance and monitoring of individual actions would greatly expand. Tighter control and surveillance including police access to all person-related data-banks would be provided with an effective legitimation.

The problems generated by a "plutonium economy" are of relevance here as they best mark the point of convergence of the need for containing technological risks, or bureaucratic development, and of the use of information technology in support of both. The need to grant greater social stability will be a powerful stimulus for the continuing expansion of bureaucratic power.

In the past, bureaucratic growth was related to the rationalisation of production, of administration and of financial redistribution for purposes of social welfare. The human frontiers of such a rationalisation and centralisation are now more and more visible. This finds its expression in exhortations to "debureaucratise" large organisations. But bureaucracies continue to gain importance. The tightening and centralisation of social control is now taking over as the motor of bureaucratic growth. Its legitimacy will hardly be questioned as long as it is concomitant with an apparent centralisation and debureaucratisation, as well as with more human autonomy and choice in domains that present no threat to the survival of the society.

In this light, there seems to be no escape from the bureaucratisation of society. The social vulnerability due to technological risks leaves no choice. Information technology, especially computing, provides the proper technology for the stabilisation of society and the expansion of bureaucratic power.

So far, the effects of information technology have undoubtedly been an increased efficiency of large organisations, as well as enhanced opportunities to monitor the behaviour of individuals and of social systems both inside and outside the organisation. To this can be added the growing ability to anticipate or predict the future behaviour of individuals or social systems which makes it easier to manipulate them.

Indeed the image of a future "organic" society may not be too far-fetched. In such a society, the co-ordination and control of individual actions is afforded by what might be called an external nervous system, with sensors and monitors giving constant impulses to head-quarters of control. The increased steering capacity thereby generated will not be broken by the decentralisation of monitoring and intervention facilities, unless the information links to the central bureaucracies are cut.

It is perhaps this increased steering capacity of bureaucracies which constitutes the main effect of the combined force of bureaucratic and computational rationality. A better anticipation of human behaviour can flow from new analytical tools for processing information, as well as from new information sources including opinion polling and electronic voting. The dangers of predicting and shaping future events include dangers to democracy that may well outweight those resulting from open threats or from surveillance of political behaviour.

Once again, what is at stake is not so much the use of information technology as such but its use by powerful groups in society, by the large bureaucracies that already dominate much of the present societies regardless of their government systems. The technology itself as well as its different applications are profoundly shaped by the desire of powerful groups to use the technology for staying in power and further strengthening their own position. Thus the tendency of power toward occupying an ever larger place in society will shape the use of the technology.

It is all too often believed that cheap home computers and the emergence of information utilities will counteract this tendency. Changes in the distribution of power are often attributed to the economies of computing. The economic power of large organisations makes them profit more from computers than powerless individuals. On such a basis it is then speculated that power gains of organisations are of a transient nature, giving way, as soon as the technology diffuses into private households, to a re-established power balance or even to new unprecedented opportunities for citizen participation in public affairs or self-fulfilment. Such speculations are unfounded. To take an example from broadcasting: the availability of Citizen Band radio, a truly interactive technology, in no sense encroaches upon the traditional domain and impact of the mass media.

Perhaps the greatest danger lies in the fact that there will be considerably less opportunity for countervailing forces to appear on the political scene and to articulate themselves. So far, bureaucratic administration has centred on economic questions including social security. It has been aimed at guaranteeing basic rules of civic and market behaviour. With better information technology aids it can now expand to virtually all aspects of social life, monitoring, predicting and shaping human behaviour in sectors hitherto considered as private and exempt from public intervention. Starting from educational institutions on one side, law enforcement and personalised social care on the other, it slowly pervades the very process of socialisation of individuals, their findings of an identity. The eventual outcome would be a perfect administration completely prestructuring and monitoring human behaviour.

No doubt a strong trend towards a more orderly, well regulated society exists throughout the world. Since Max Weber's early analysis of an imminent bureaucratisation of society this is a recurrent theme in Western thought. Weber's anxious question of whether, as a result of the rational bureaucratic management of all important human affairs, we might become powerless inmates of a "cage of serfdom" (*Gehäuse der Hörigkeit*)[10] is much more pressing now than it was some sixty years ago when he feared that socialism would lead to bureaucratic structures stifling innovation and social evolution.

Still, there is no "hard" technological determinism at work. The use of information technology is shaped by social forces. Therefore, the outcome is still open. Information technology does not determine, in the strict sense of the word, the coming of a society dominated by bureaucracies and functioning like a "megamachine" (Lewis Mumford). But it makes it more difficult to oppose it.

[10] Max Weber, *Wirtschaft und Gesellschaft*, pp. 1059 ff. Studienausgabe, Köln, 1964.

10

Microelectronics and World Interdependence

ALEXANDER KING

At the outset it was pointed out that microelectronics, arising as it does from scientific research and technological development in the laboratories of a few countries of the First World and immediately applicable to the industrial and service activities of these countries, has provoked initial concern with regard to its impact on the industries of the North. There is, as yet, very incomplete comprehension of what may be the ultimate significance of microelectronics for society, and certainly this will be a matter of controversy for many years to come. Nevertheless, we can be certain already that the new technology with its wide span of application will create a major discontinuity in both the industrial development of the North and the economic growth of the South. Its influence will be global and it is unlikely that its impact will by-pass any type of society. We return, therefore, in this chapter to a discussion of some of the political aspects of the development, especially, but not exclusively, related to its possible influence on the international distribution of industry.

Microelectronics and Industrialisation

Three aspects have to be distinguished: (i) the growth and potentiality of the microelectronics industry as such, including its important software elements; (ii) the development of new products or the adaptation of existing ones to take advantage of integrated circuit technology; and (iii) the full use of such devices, domestic or imported in the automation of industry and services including automatic control

311

and integrated functioning. The normal diffusion of technology across national frontiers, to which we have already referred, will in any case make the applications of microelectronics available to all countries, quickly in the case of those already possessing advanced industries and more slowly for those at an earlier stage of development. However, those countries which have created their own facilities for the manufacture of silicon chips and integrated circuits and also possess a sound software capacity will, through intimate familiarity with the techniques, have a considerable advantage in using their applications quickly and effectively, over those which have to buy the "packaged" technology.

The introduction of manufacturing processes based on microelectronic circuitry, in which a high degree of artificial intelligence is embodied, will require sophisticated planning techniques and will call for the training of people with a variety of new skills; quite new systems of management will be needed and there may be considerable amounts of industry located away from traditional areas, accentuating the trend of recent decades to locate advanced manufactures near to important centres of research and training rather than was the case at the time of the Industrial Revolution, when industries were placed close to reserves of mineral ores and coal or near to estuaries from which their heavy products could be shipped.

It must be remembered also that the existing distinction between the traditional industries and those depending on advanced technology and research will tend to disappear as microelectronics causes the existing traditional manufactures to become skill-intensive with the intelligence now incorporated and the systems knowledge necessary for their successful operation. While such modifications of industrial structure are rather obvious in the case of the highly industrialised countries, similar criteria are likely to be followed, although more slowly, in relation to Third World industrialisation, at least in those sectors which aim at finding a substantial role in the world market. There will, of course, be many instances where existing types of process are still appropriate to meet the developing countries' needs, especially with regard to import substitution and the manufacture of goods for local consumption where transportation costs, etc., will operate against foreign imports.

We have suggested that many of the social consequences of the microprocessor and of automation may be somewhat long term in their effects, but that in the earlier stages conflict may arise between the international competitive advantages of the new technologies and structural employment which they will generate. This is, however, unlikely to inhibit the introduction of the new technologies especially in the industrialised countries poor in minerals and energy, for which technological innovation, high added value of their products and high productivity will be desirable if they are to maintain their economic positions. While we feel that eventually quite radical approaches to employment and unemployment, involving deep social change, will be necessary during the next few years, it is probable that more traditional approaches will be taken. Recent decades have seen a happy combination of rapid technical change, rapid economic growth and cheap energy in which industrialised countries were producing for expanding world markets. It is highly improbable that such a favourable economic climate will exist in the near future. More likely, in a period of slower growth and greater constraints, the present highly industrialised countries will be competing with one another for a share in a more sophisticated but possibly more restricted market. Such prospects could be serious for global employment, but they likewise suggest that a Luddite-type resistance to the introduction of microelectronics-derived processes could be counterproductive and will be seen as such by the workers and their unions.

Nevertheless, some serious upheavals are to be expected in industries where the introduction of automated processes leads to a sudden or high rate of redundancy. Indeed, bitter disputes have already arisen with regard to computerised operations in certain industries, such as printing in several countries, but such difficulties can be minimised if there is sufficient advanced warning, intelligent consultation and facilities for training the displaced workers when the new technologies are introduced. The real difficulties will arise if and when there are no obvious alternative employment prospects for which to retrain.

An important aspect of this problem is to involve the trade unions in a general as well as a sectoral sense in consideration of the difficulties. This has already been begun in some countries, an early example being the 1975 agreement between the Norwegian Federations of Employers and of Trade Unions, which required full consultation between management and unions before computerised processes are introduced. It is becoming ever more clearly recognised that for some highly competitive industries, problems of redundancy and reassimilation of displaced workers have to be faced up to if there is to be any stable employment situation at all. The inescapable employment aspects of the application of the new microelectronics should be turned into an opportunity to achieve a greater degree of industrial democracy.

The above and other considerations indicate that a mastery of the possibilities of microelectronics must be an objective of all countries. The rate of application may well be slowed down by difficult economic circumstances in general, but such circumstances make the new developments seem to be both a threat and an incentive. International competitive conditions will determine the tempo of introduction, and no individual countries will be able intelligently to choose their own rates; this is a truly global situation, in which, as we shall argue later, the United States and probably even more Japan will be the pacemakers.

Evolution of the Microelectronics Industry

The distinction between the microelectrics industry as such and the general transformation of the industrial and other sectors which it will encourage must be stressed once more. In this section we deal with the industry itself. Nearly all the early innovations were made in the United States, and American companies still account for approximately 70 per cent of the global production of integrated circuits, but this situation is changing with the rapid technological advances of the Japanese companies and to a lesser extent of Europe, so that American dominance is threatened. The American effort has been greatly assisted by substantial research and development contracts and other financial support from the Federal Government through the Department of Defense and NASA, which also provided a guaranteed market for the first integrated circuits, and this enabled the American companies to move quickly into the production of new semiconductor devices before a commercial market was established. A further advantage was the dominant position of IBM in the computer field, which quickly made it a major customer for the new circuits. While the American defence requirements account today for quite a small proportion of the integrated circuit market, defence contracts still give a major incentive to advanced research in the field. For example, Pentagon support at the level of $200 million is being devoted to an industrial effort to develop very-high-speed integrated circuits (VHICs) which will contain hundreds of thousands of closely packed components, a development which will have obvious repercussions in civil application.

The Japanese path to microelectronics development has been necessarily different in the absence of a substantial defence research effort. The success of the Japanese traditional electronics industry, centred on a few powerful corporations, made these enterprises fully aware of the significance of the new possibilities and of the American advances. This country, more than any other, took an early comprehensive view of the possibilities and was convinced that an electronic-dominated industrial economy was over the horizon. About fifteen years ago, therefore, a deliberate policy was agreed for the establishment of the "information society" and at the same time a government sponsored effort was initiated to promote a strong, domestic computer industry, protective measures being taken at the same time through MITI (the Ministry of International Trade and Industry) to shield the domestic industry from foreign competition. One of the elements of the Japanese programme was a joint government-industry research project to develop the necessary techniques for high-density integrated circuits, to which the government contributed $250 million, or 40 per cent of the total. This effort has succeeded well and the Japanese are now not only providing stiff competition to the American firms, but have a leading position in some of the current technological developments. One aspect of the Japanese situation is that their microelectronics enterprises are units within large electric and electronic companies, in contrast to the American structure in which the initial innovation came from a small number of small enterprising firms which started with quite modest amounts of investment capital in "Silicon Valley". Thus the Japanese projects were in favourable position to raise capital, an important advantage when one realises that new production lines for the manufacture of advanced circuits may cost as much as $40 million. It is not surprising that in recent years several of the successful American semiconductor firms have been taken over by large corporations. This is causing some concern in the United States, since it is feared that by becoming elements within large corporations, these innovative firms which were able to bring each technological breakthrough into rapid commercial exploitation, thus beating the traditional electronics giants, may lose their flexibility in contrast to the Japanese experience.

European industry, lacking the markets for space exploration and defence, which made possible the rapid microelectronic development of the United States, and without the combined government-industry strategic approach of the Japanese, has moved more slowly into the field. In 1979, for instance, only one third of the integrated circuits used by European industry were manufactured by European companies, most of the rest coming from the United States. By that time, however, the importance of microelectronics for the industrial future was realised by the main industrial countries of Europe and, as a consequence, considerable but tardy efforts were made to ensure a place in this key development. In the United Kingdom, for example, the government has invested about $120 million in a company called Inmos for the manufacture of the next generation of memory chips. The hope here is that by establishing an advanced integrated circuit industry, not only will there be an expanding market for its products, but that it will stimulate the interest of potential users and thus assist in the modernisation of industry in general. Analogous efforts to stimulate research and development in the field and to stimulate the creation and growth of semiconductor firms are being made by the governments of the Federal Republic of Germany and France.

While there is vigorous competition between the main industrial countries in the manufacture and marketing of integrated circuits, the future prizes are likely to be much greater in the application of such devices in products and processes for the automation of industry and of tertiary sector activities. At present the United States has a dominant position in the use of integrated circuits as the following table indicates.

Distribution of Semiconductor Device Use among OECD Members, 1965-1978 (in per cent)

	1965	1970	1974	1978
United States	66	53	48	46
Western Europe	18	22	25	23
Asia (mainly Japan)	14	22	24	28
Other	2	3	3	3

Source: Charles River Associates, *Innovation, Competition and Government Policy.*

However, Japan is moving rapidly to introduce microelectronics technology into as wide a range of industries as possible, and European countries such as France, the Federal Republic of Germany, Italy and the United Kingdom are all making strong efforts to deepen the understanding of their industrialists as to the significance of integrated circuit technology for their competitive prospects and to assist in the application of the new devices. In total, European governments are spending about $1 billion on such efforts and private industry is investing about the same amount. For perspective, however, it should be realised that in the United States one single firm, IBM, spends about $1 billion per annum on research. The European countries will find this development difficult, as a result of their late arrival in this field. Nevertheless, the race is on, a race which is a response to international economic pressures which force most of the industrialised countries to compete, although they have little say in its rules or its course. Still more disturbing, however, is that the efforts of all the competing countries to stimulate the technology as rapidly as possible are not being matched by policies or even thorough investigations as to the social consequences and negative side-effects—notably the inevitable conflict between productivity and employment.

Reflection on the probable influence of the social factors in microelectronics development and how the various nations are likely to deal with it leads us to the conclusion that Japan is likely to be the pacemaker. Certainly Japan has a strong technological position and a coherent medium- and long-term policy for microelectronic industrialisation; already at least half of the world population of industrial robots is in that country. However, in the long run it may well be that the nature of Japanese society, the behaviour of its institutions, its industrial and human structures and its genius for reaching decisions through consensus will prove to be even more important.

Deep consideration by the governments in Europe and North America of the social consequences of the new and dominant technology which they are stimulating as vigorously as they can would seem as important in preparing the basis for innovative policy to meet the new situation as the technological stimulus itself.

Microelectronics and the North-South Disparities — Risks and Opportunities

Earlier in this book there has been a somewhat detailed discussion of how microelectronic technology is likely to affect the prospects of the countries of the Third World. Here there is need only to repeat some of the main conclusions in relation to the overall international situation. There are two main effects to be anticipated. Firstly, and fundamentally, the automation of factories in many industrial sectors of the advanced countries through microelectronics-controlled manufacture and computer-directed assembly will erode the main comparative advantage of the developing countries, namely their low labour costs. This is particularly serious at the present time, because the world-wide slowing down of economic growth will tend to encourage enterprises in the developed nations to use the new automation possibilities to revitalise industries such as textiles and garment-making, which in the expensive conditions of the fifties and sixties were allowed to yield to the cheap labour countries of South-East Asia.

The second effect is that the concentration of a new and vital technology in the main industrialised countries, with massive efforts of research and development in support and with its direct stimulus to a wide spread of economic sectors, is likely to increase the gap between the rich and the poor countries. The microelectronics revolution may well result in a quantum jump in the degree of sophistication of world industry, rooted firmly in the countries having the necessary scientific and technological competence. But reciprocally, microelectronics in enhancing this already considerable competence will display all too starkly the relative impotence for industrial development of those countries whose science and technology infrastructure lies well below the critical theshold. This is a classical example of the Biblical saying, "Whosoever hath, to him shall be given...". Here, of course, we must heed the warning, given earlier, not to generalise with regard to development; there are certainly a few countries in the group designated as of the Third World, such as India, which possess a considerable degree of scientific skill and awareness to take part in the new development; nevertheless, the great majority of the developing countries are simply not in the race. Computerised and automated processes will of course gradually diffuse into such economies, but much later than into presently industrialised nations. The most effective mechanism for technological transfer at present is the transnational corporation, despite the suspicions which their operations generate in Third World countries. This is not the place to discuss the validity of the criticisms which are levelled against the transnationals, a powerful group of enterprises with as great a diversity of corporate behaviour, policies and motivations as the heterogeneous grouping of the developing countries themselves. Undoubtedly many of these corporations are attempting to establish, in the interest of their own long-term survival and prosperity, a sense of identification with the objectives of countries within which they operate. It is unlikely that any alternative and effective means of large-scale technological transfer will be operative in the near future, and hence the transnationals, although not loved in the developing countries, are needed and will be the main agents for the introduction of microelectronic products and systems.

Much of such transfer is likely to be in the form of complete "packages" of technology; these may well help to create wealth, but certainly not employment, which is one of the main problems in most developing countries and one which will be acutely aggravated by population increase. In such countries microelectronics tends to displace low-skilled labour in favour of smaller numbers of high-skilled workers, thus increasing the difficulties of many countries both in absorbing the low skills and training the high skills. It is suggested in some quarters that the rapid introduction of packaged microelectronic processes to the developing countries could be the panacea for the solution of the North-South problem, allowing these nations to leapfrog into the sophisticated industrial world of the twenty-first century. This approach is rejected as unrealistic, misleading and diversionary. Undoubtedly, microelectronics *can* help in the development of the Third World, but its wise use will be a slow, subtle and costly process, involving many social and cultural as well as economic factors. However, the real problem is not that too rapid introduction of the new technologies will generate socio-cultural difficulties, but rather that they will be insufficiently taken into consideration in development strategies. It is significant that the promise and problems of microelectronics application and other technologies have hardly been discussed in recent conferences on the industrialisation of the Third World. It would be a tragedy if these new trends were to be ignored, resulting in many countries' industrialising to obsolescence.

One matter of strong psychopolitical importance should be mentioned at this point. The acceptance of northern advanced technologies by the South—not only semiconductor development, but also nuclear technology, modern information and communication methods including the use of satellites, petroleum refinery and many others—will greatly increase the dependence of these countries on the North and on its transnational corporations, will make the emergence of indigenous technologies more difficult and, in the end, will erode local cultural values. This will be seen by many in the Third World as a new wave of technological colonialism.

For the Third World, the present technological situation is critical and there is a desperate need for the reassessment of development strategies and the search for new approaches which will take into account the promises and threats of the new technology. As one representative of the Third World, participating in a discussion of the first draft of this book put it: "There must be a clarion call to examine the two available options, either to maintain the status quo and to see the gap between developed and developing countries widen or to seize the revolutionary possibilities of the new technology and exploit them quickly within a new development strategy. We must not wait to see what the First World will do and then imitate it. If we miss the present opportunity, it will not recur.''

The spread of the microprocessor, the computer and of automation may indeed prove to be a critical point in world development with many social consequences. The techniques which emerged from the first industrial revolution established a firm path to development of an essentially material nature, from which the northern countries, with the puritan work ethic, greatly profited and which has insidiously spread throughout the world, threatening many indigenous cultures which have been insufficiently strong to resist it. In recent years there has been a growing recognition of the value of cultural diversity, not only for its own sake but for sound biological reasons. Thus many voices have been raised in different cultural and religious regions, of which that of Gandhi has had special appeal, suggesting the need to seek other life styles than the materialistic way of the North. Even in the industrialised countries many people, and especially the young, sensing their alienation from the technology-dominated civilisation in which they live, are attempting to find new life styles and establish counter-cultures. In some developing countries also, many thinking people who accept the need for technological progress in the raising of material standards to eradicate poverty and hunger, would like to find their own technological paths to development and particularly to encourage the evolution of labour-intensive processes which will provide jobs as well as wealth. However, the acquisition of material benefits seems to have a general appeal, despite cultural trends which may counter it, and in nearly all developing countries the élites seem to have pretty well the same appetites as people in the rich countries and lean towards conspicuous consumption in the form, for example, of luxurious automobiles and high animal-protein diets. A large number of people in such countries are also attracted by the gadgetry of the industrialised North. The new level of sophistication of industry and its products which microelectronics is bringing about is likely to contrast even more strongly with local goods, and when it spreads to the developing countries the cultural dichotomy will be stark and probably unbridgeable.

So far we have considered mainly the difficulties which the developments of microelectronics will pose for the developing countries, but the effect will be by no means completely negative. One overriding consideration is that the new information technology should give access to the totality or near totality of the world's knowledge through contact with an international range of data banks and information services. The limitations here will be the capacity of developing countries to select from the enormous mass of facts and ideas those items which are specifically significant to each country in question. Here we must repeat our monotonous insistence that the critical factor in each case is a critical capacity for science and technology and for management. Beyond this, many specific opportunities for the application of microelectronics exist in the developing countries and these will increase as costs are reduced. There are, for instance, many possibilities for the computerised control of systems which will come within the means of communities and individuals as the price of minicomputers continues to fall, and this can provide a considerable increase in efficiency in many activities such as river basin management, irrigation control, remote sensing, urban system problems, education and rural systems optimisation. The village computer may well become commonplace after a number of years in assisting in the distribution of effort to ensure that the available biomass is fully used and sustainable for the production of both food and energy. Nor can one rule out the widespread use of the computer in agriculture. For example, an interesting project is in operation in Venezuela at present whereby local climatic and other conditions fed into local minicomputers are analysed centrally and give information as to the right moment to plant, to apply fertilisers, to spray against pests and fungi and to harvest. Such an approach is as yet insufficiently tested out in real conditions, but it gives promise of increasing yield to an extent which will far outstrip the extra costs.

Microelectronics and Industry in Countries
of Different Types

It is generally accepted that industrialisation in the Third World is the key to further economic development. Despite notable progress in the newly industrialised countries of South East Asia and a few larger nations such as Brazil and Mexico which are aspiring upwards from underdevelopment, the proportion of world industrial production outside the First and Second Worlds remains very small. The United Nations has established optimistic targets for overall industrialisation in the Third World, but these are hardly realistic without worldwide agreement on changes in the international division of labour. This would necessitate considerable increase in the quantity and value of manufactured goods imported by the industrialised countries from the less developed ones, with adverse effect on a number of their own more traditional industries. Even in recent years of high economic growth, the depressing effect of massive imports from Asian countries on the textile and garment sectors of the United States and some European countries has led to some resentment in the industries affected, and it will obviously be still more difficult in the present climate of economic difficulty. For the same reason, the North-South discussions are in impasse and little progress is being made towards the establishment of the New International Economic Order which has been discussed at length in the United Nations and which seeks to establish a world economic system of greater equity between the rich and the poor nations. The incursion of microelectronics increases the uncertainties, but, nevertheless progress is being made in many directions. It is highly probable that oil refining will be undertaken progressively by the petroleum-producing countries themselves, and that the petrochemical industry will be effectively coupled with such development. Where this is so, highly integrated plants are likely to be introduced, with microelectronic control of the process. Most of such development will of course be initiated with the help of transnational corporations, but it will be a situation propitious for the training and experience gaining of young scientists and engineers in the developing countries concerned and eventually for the diffusion of microelectronic techniques to other industries and systems.

In general, high costs of energy should encourage a greater proportion of the exported natural resources of the Third World being processed as near as possible to their points of origin. This should be the case particularly for the beneficiation of minerals and would be especially favourable in cases where fossil fuels or hydroelectric power is available. After all it is surely uneconomic to haul mountains of ores thousands of kilometres to the industrial user countries, because of the small amount of metal they contain. A number of "intermediate countries" now rapidly industrialising will no doubt continue the process with success. India is a particularly good case of a country manufacturing a wide range of equipment such as machine tools and possessing a well-educated and extensive scientific and technical manpower which should be able to profit by the developments in microelectronics, although at a slower pace than the older industrialised countries. Mexico and Brazil may also be approaching the critical threshold of scientific and technological competence and, with their high natural resource potential, should have a bright future and, if they succeed in evolving integrated economic and technological policies, should again be able to develop modern industries using microelectronic devices. All these three countries have a high potential in software.

The oil-producing countries as a group are in a somewhat different position. They possess in their production, exploration and refinery activities an extremely high technology sector, but one introduced in most cases from abroad, by the oil transnationals, with very little other industrialisation and, as yet, very marginal capacities for research and development. In the case of the Islamic countries, the cultural compatibility of technology may raise many difficult problems as the case of Iran has demonstrated. However, recognition of the limited life of their major resource and with the benefit of the high level of capital accumulation which their petroleum has provided, they are likely to attempt a much wider industrial development, in which integrated circuit processes could well play an important role, the new technology being largely imported in "packages" with the help of the transnationals. Another important element in the situation of the Arab oil countries, other than Algeria, is their very small populations. Even today very many foreign workers have been employed and a more widespread industrialisation would necessitate a considerable increase in their numbers, with potential political problems. For such countries a major effort in education is required and the building up of a research and development capacity. Kuwait is making considerable efforts in this direction as is another OPEC country, Venezuela, but in general scientific and technological efforts in the Arab oil countries are marginal.

We come next to the special problems which integrated circuit technology brings to the "newly industrialised" countries of South East Asia. These nations are likely to be affected immediately, because their export success has been mainly in manufactures where cheap labour has provided the competitive edge over similar products from the industrialised countries, namely textiles, garments and the offshore assembly of electronic goods on behalf of the multinational corporations. The probable impact of the microprocessor has already been discussed in some detail in previous chapters, and here we shall only recapitulate a few of the trends within the perspective of the international distribution of labour. Textiles, garment-making, shoes and leather goods account for more than 40 per cent of the exports of manufactured goods from Hong Kong, Singapore, South Korea and Taiwan, and the growth rates in the volume of these exports increased dramatically during the sixties and the early seventies, although these rates of increase have slowed down considerably in recent years and a more modest increase of 5 per cent per annum is projected for the eighties—if the impact of microelectronics is not too brutal. In industrialised countries, the prospective gains from the automation of these labour-intensive industries have already tempted entrepreneurs in the traditional industrialised countries to transform their processes. The American textile industry is at present expected to spend about $2 billion per year in the eighties on advanced equipment, much of it computer controlled, in the expectation of increasing its share of the world market, with a reduction of its labour force of about 300,000. As the effectiveness of integrated circuit technology becomes more widely recognised and textile machinery makers become more experienced in its applications, the rate of automation in this industry may well increase more rapidly than at present forecast. The European textile industries are also alert to the new possibilities; a British spinning mill, recently partially automated, claims that its new plant, operating with expensive equipment and with a work force of only 95 people, will produce as much as three obsolete plants with 435 operatives. Much the same situation is evolving in the garment sector, although this will be considerably slower than in textile manufacture, owing to the inherent complexity, number and variety of the operations. In these and other sectors, automation in the North is bound to erode the comparative advantage enjoyed at present by the low labour cost countries of the South.

The situation with regard to electronic assembly in the "newly industrialised" countries is somewhat different from that of the sectors just mentioned and is even more directly related to the electronic developments of the United States, Japan and Europe. It is relatively unconnected with any indigenous industrial background in the countries concerned, all of which have long experience in the textile industry, for example. The existing situation is essentially one of the advanced industrial countries, through the transnational corporations, exploiting the cheap labour and manual dexterity which exists plentifully in the Asian countries concerned. Electronic components are shipped in to Singapore, Taiwan, etc., where they are soldered together to larger assemblies and finally mounted to form the final product which is then re-imported to the country of origin of the components or their export customers.

Such offshore assembly has brought considerable earnings to Asian and some Latin American countries, but it has some questionable features. It employs many low-paid, low-skilled workers in highly regimented and repetitive work, involving hours of manipulation done while peering through microscopes. Such work is not only costly when undertaken in the northern countries, but is subject to much stricter work and health regulation. The job creation resulting from such arrangements is likely to slow down considerably and may eventually disappear. The electronic companies which profit from it will obviously be amongst the first to apply the integrated circuit to automate their own industry. The tendency is, therefore, for the soldering and testing of integrated circuits as well as final assembly to be undertaken automatically in the domestic plants as the real labour costs diminish. Such highly automated domestic assembly is likely to become a major determinant of international competition, and even now investment in new plants is being made in the industrialised rather than the developing countries. Recent reports from Japan, for instance, indicate that through a degree of automation in the manufacture of television sets, an increase in production has been achieved, with a considerable drop in the work force.

Its seems probable that the newly industrialised countries of Asia will be the first group in the Third World to feel the impact of microelectronics, but, at the same time, they will be amongst the most capable of coping with it. Much adjustment will be necessary in the economic structures of these countries, and times are not propitious for a smooth transition—the general condition of the world economy, high energy costs, widespread inflation, protectionist tendencies in the North and changing consumer habits—all these will play their part. A special feature in this part of the world is the nearness to the upsurge of Japanese technology. Japan is already making important technological incursions into the Third World and especially into the newly industrialising countries which are near to her, and Japanese microprocessors and industrial robots are likely to be more important than the packaged technology of the Western transnationals. However, these countries have profited greatly from their experiences of the last two decades and have learnt much concerning industrial strategy, the operation of modern industrial plants, management techniques and international marketing. Furthermore, they possess a strong sense of entrepreneurialism. While they have by no means reached the critical threshold of scientific and technological competence, they understand the need for it and have already established a certain infrastructure for such activities, which has kept them informed of the significance of recent scientific discoveries wherever they may originate. In the case of Singapore, for instance, there is a quite sophisticated scientific awareness and there are research organs with excellent relations with the corresponding bodies in the North. Above all, such countries appreciate the importance of education and training, both general and technical.

The various categories of countries discussed above include less than half of those normally bunched together with the label "Third World" For the remainder, the prospects offered by the development of microelectronics are distant, indirect and, in a relative sense, unfavourable. There will indeed be benefits for them, by-products of general world development, in information accessibility, communications and transportation and hopefully also in the development of new labour-intensive technologies using advanced scientific approaches, but not in the traditional way of replacing men by machines. The high level of solar energy reaching tropical countries gives hope that they will be able to develop new means of using this radiation, which is the one external input which prevents our planet from being a closed system. Recent advances in biology offer real, but probably no immediate possibilities for, improvements in the least developed countries. Developments arising from the new genetic engineering techniques and advances in enzyme technology could be the basis for energy generation and the integrated use of the available biomass and could possibly give rise to important chemical industries. These do not of course necessarily involve microelectronics, but if their potentialities are actualised, microelectronic controls and integrated systems will almost certainly be incorporated.

The advent of the microprocessor and the prospects of a rapid advance to the automation of the industries of the North may well, at least in the early years of the development, widen the gap between the rich and the poor nations still further. It is essential, therefore, that there be a search for new development strategies and that these be evolved within the perspective of total world development and take into account the potentialities and threats of microelectronics and other impending developments of technology. For the least developed half of the world's nations, the following desiderata would seem to be amongst those of the highest priority:

(i) In view of world technological and industrial trends, outlined in this book, there is an urgent need for such countries to reassess their policies, industrial and otherwise, and to search for new paths to development which will make full use of the new possibilities and avoid their dangers as far as possible.

(ii) They should seek to create a critical level of awareness of and effort in science and technology, without which substantial indigenous development is improbable.

(iii) In pursuance of (ii) they should endeavour to create regional groupings which would have a much greater probability of achieving the critical threshold in the not too distant future than would the efforts of the individual members of the grouping separately.

(iv) Once formed, these groupings should seek to forge links between themselves and also with friendly bodies within the industrialised countries; these linkages might take the form of mutual interdependence treaties for co-operation on development strategies, scientific and technological reinforcement and allocation of tasks, access to data banks and information pools and the training of technological and managerial skills.

For Better or for Worse?

Microelectronics emerges as a new force in a world, if not in crisis, at least unsure of itself, its values, its goals and its destiny. We live under the Damocles' sword of nuclear annihilation, surrounded by many "little", conventional wars, in economic difficulties, creeping desertification and other forms of environmental deterioration which threaten health and climate, population explosion, increasing violence and alienation of individuals from society. The basic question for humanity is whether this new force, while possibly increasing the material prosperity of some, will be allowed to aggravate this situation still further or whether we can generate the wisdom to use it positively to shape new forms of society, with greater equity, which can offer a life of dignity and modest prosperity to all people and opportunities of human fulfilment? We have no hesitation in proclaiming the possibility that the latter option can be achieved, although the transformation would entail some sacrifice to the privileged minority.

The argument of the book is that the development of microelectronics and its probable widespread application in economic and other fields will constitute a major political force within the next few decades, with considerable influence on the international distribution of labour and eventually with impact on all countries, irrespective of their ideological and cultural traditions. However, technology in general and microelectronics in particular must not be considered in isolation, as an autonomous force, but looked at in its reciprocal relationships with other forces, trends and situations. Until now national plans, including the development plans of Third World countries, have been dominated by macroeconomic thinking and technological factors have usually been ignored. The impact strength of the new technology is such that this can no longer be accepted. A number of the other factors with which microelectronic developments will react are associated with the particular difficulties of this period: slow growth, including high energy costs, a swing towards protectionism on international markets, restrictive monetary policies to combat inflation, and general structural unemployment. The rate of introduction of microelectronics and its applications may well be slowed down by these difficulties; on the other hand, many countries, favourably placed with regard to this development, will inevitably make great efforts to advance it as rapidly as possible, so as to achieve a substantial share of the world market. Those which do not develop and apply the technology sufficiently rapidly will see their competitive position and share in international trade deteriorate. While microelectronics will certainly help to stimulate growth in some areas, it is no panacea for general economic recovery.

Thus one immediate effect, already becoming evident, is an increase in the intensity of international competition between the most highly industrialised countries. The United States is at present in a leading position in this race, because of its initial innovations in integrated circuitry, but Japan, as we have suggested, for technological, industrial, structural and cultural reasons, is likely to become the pacemaker, especially in applying the new technology quickly and comprehensively across a wide spectrum of economic sectors; the European countries are at present lagging behind, but are fully aware of the problems and prospects and are now making strenuous efforts to catch up. Some of the other industrialised countries remain virtually underdeveloped with regard to microelectronics.

However, it is with regard to the possible social consequences of microelectronics that there seems to be most ground for concern. Its possible influence on employment, for example, is understood, but measures suggested to meet what may possibly be an endemic, high rate of unemployment are those which were evolved for simpler societies, now becoming obsolete, and there is little evidence that any of the countries now suffering high levels of unemployment are having any marked success. This book suggests, therefore, that more radical means are required, and appeals to governments, trade unions and employers' organisations to study the employment consequences of microelectronics more deeply, so as to provide a basis, both for contingency measures, should they be required, as well as longer term and more fundamental approaches which will have strong societal impact. Some suggestions have been made here as to possible directions to take, but there is need for comprehensive research in this field. In the past there has been little feedback influence on the direction of technological development from society, but in recent years, with the arising of the environmental and conservationist movements, soft energy lobbies, nuclear refusal and opposition to supersonic aircraft, there is ample evidence of serious public concern about the use and abuse of technology, and governments would be well advised to give deep consideration to the social consequences of microelectronics before they arise in an acute form. Otherwise they may be surprised by opposition which may appear fanatical and uninformed, but based on instinctive concern.

This raises the question of the adequacy of governmental policies, structures and attitudes to deal with the problems likely to be posed by microelectronics. The structures of governments were created for earlier, simpler times and despite the great increase in official functions since the end of the Second World War they have only been modified marginally. One major shortcoming in most countries is the absence of sufficient or sufficiently accepted machinery for overall policy co-ordination and planning. This is of course formally the responsibility of prime ministers and their Cabinets, but in practice the mass of decisions required and the heavy preoccupation of heads of governments with internal and external political issues is such that too little attention is given to the integration of policies. Nearly all the great problems of the day are horizontal in character and sprawl across the whole spectrum of governmental function. Governments, on the other hand, are organised vertically through ministries for specific sectors. As a result, national policy tends to be the summation rather than the integration of sectoral policies, some of which reinforce each other, of course, but others may be in conflict to a greater or lesser extent, recognised or not. The energy situation with its problems impinging on so many elements of the machinery of government has shown up this weakness clearly and many governments are trying to find means to overcome it. The development of microelectronics together with its applications and the probability of profound social and cultural consequence equally require an integrated approach.

A further matter for which institutional innovation will be necessary if microelectronics is to be used deliberately and constructively for the improvement of society is to ensure that a sufficiently long-term perspective is taken in preparing for changes which are inevitable but may not always seem pressing. It is not easy for governments, especially within the democratic system with an electoral cycle of about four years, to face up to long-term problems. Such issues, no matter how fundamental, tend to be crowded out by short-term issues which interest the electorate, no matter how trivial they may be. The probable impact of microelectronics is so great and its probable societal consequences so profound, especially when seen in relation to the other elements of the great transition which was described in the Introduction, that policies are needed in terms of decades rather than of years if it is to achieve the benefits which are possible for humanity as a whole. The normal, short-term perspective of governments will tend to face these problems by a series of policies of expediency to meet the issues as they become inescapable and thus attempt remedial action and marginal adjustments *post facto,* rather than forward-looking measures aimed at shaping and steering the development to the maximum common good.

We return now to the international issues and especially the North-South issue and the need for universal recognition of the reality of the interdependence of the nations which the new wave of technological change will render ever more apparent. The development and application of microelectronics adds a new and important factor to international competitiveness, which can be exploited initially only by the scientifically most advanced countries. This will tend to widen still further the technological and hence the economic gap between the advanced and the underdeveloped nations, especially as it reduces one of the few advantages of the latter, namely the availability of low-cost labour. In fact the information breakthrough due to microelectronic devices and systems should greatly enhance the research capacity of the scientifically advanced countries and only slowly that of the less developed, lying well below the critical threshold of scientific awareness and competence. Thus there is danger that the attempts of the developing countries to introduce microelectronic methods will still further increase their dependence on the industrial giants of the North. A decolonised world dominated by what will appear to be scientific imperialism makes little sense.

This, as well as many other factors including the growing disparity in population numbers between the rich and the poor countries, suggests the need for a new look at the North-South issue. At present, discussions on this fundamental world issue are in impasse; recent discussions indicate that it is a dialogue of the deaf. To many leaders in the North and, indeed, to much of the general public, development aid appears to be based on humanitarian charity, mixed with a good sprinkling of self-interest in the form of commercial benefits including the sale of arms, political influence or cultural expansion. To many in the South, the donor countries are, or should be, motivated by the guilt of past colonialism, political or economic, and aid is seen or is said to be seen, as a right and, in fact, a quite insufficient reparation.

In most of the discussion on development aid, the problem is seen from a strictly economic angle in terms of transfer of capital and more recently as transfer of technology which is another form of capital. Many of the deep structural, social and cultural issues are ignored, as the recent example of Iran has demonstrated all too clearly, and technology in the deeper sociocultural sense of its assimilability is also often ignored. It is significant that at the 1979 Conference of the United Nations on Science and Technology for Development, although the importance of the need to create indigenous capacities for science and technology was recognised, industrialisation was considered essentially in terms of a static technological situation; the consequences of the new breakthrough technologies on Third World industrialisation were hardly mentioned. Again, the recent report of Brandt Commission takes the traditional economic line and is silent on technological impact.

There thus appears to be either considerable ignorance in political circles on the significance of the new developments or an unwillingness to recognise it, and some corrective measures would seem necessary to make available the necessary information as widely and deeply as possible. There is need to provide the Third World countries with up-to-date and well-analysed data on the major world developments in science and technology and their significance for development planning. Such a service, which might be provided by the United Nations, possibly through its Development Programme, would not be restricted to microelectronics; there are many other current developments in technology which must be taken into account in national planning, but which are usually forgotten. Examples are the possibilities for the application of glass fibre technology, of great concern to copper-producing countries, or the recent decision of the American Courts allowing the patenting of new organic species developed by genetic engineering.

In discussing the way ahead, the emphasis should be on world strategy rather than merely development policy in the current sense. The raising of economic levels in the Third World and especially of the masses of the poor in these countries to whom economic benefits seem to trickle down painfully slowly cannot be tackled by them alone. It is a world problem which, particularly in view of the population explosion, will have deep consequences on countries at all stages of development. None of the industrialised countries will be able to escape, for example, the increasing demands of the developing countries for energy, minerals, food and capital. Yet, while much lip service has been paid of late to the concept of the interdependence of the nations, its reality is quite superficially appreciated by the general public in the industrialised countries. True, the recent petroleum crises have generated some realisation in these countries of their vulnerability to political events in distant lands beyond their control and hence of interdependence, but this realisation has not become generalised or its political significance fully appreciated. There is as yet little understanding in the industrialised countries of the fragility of their material prosperity, of their vulnerability to the withholding of vital imports of the materials and energy on which their industries depend, or of the consequences of the huge increases in population in countries on which they depend for imports and exports. Furthermore, governments, political parties and the public do not sufficiently appreciate their impotence in solving domestic problems which are in fact dominated by world trends outside the control of individual nations. Such a situation is inevitable in a world of some 150 sovereign states and in which sovereignty is sacrosanct. While all countries proclaim the sanctity of their sovereignty, in fact it is being slowly eroded by the necessities of international trade, international agreements necessary for the operation and management of an intrinsically interdependent world, the activities of the transnational corporations and, perhaps more importantly, by technological development. The applications of microelectronics, which we have described, cannot but accelerate the erosion. As Stanley Hoffman, the political scientist, has said: "The vessel of sovereignty is leaking."

In the probing towards a world strategy which the new technologies will make still more necessary, the key is certainly the achievement of a general understanding of the fundamental need for policies, national and international, to be based on the concept of interdependence. This necessitates a massive effort in mass education to convince people generally that a harmonious world and enduring peace can result only from the establishment of a basis of enlightened common self-interest between the nations. To achieve this will entail sacrifices on the part of some, and considerable adjustment of the pattern of industry, social values and life styles. Only after a general acceptance of interdependence will it be possible to make progress towards the establishment of a New International Economic Order. However, such a new order must not be merely economic, but must embrace all elements of human concern, including the technological.

In conclusion, it must be reiterated that a massive and all-penetrating development of microelectronics is already with us, for better or for worse. The promise it holds for human betterment and for the abolition of poverty is enormous; the degradation of society which could be the result of unwise exploitation of applications could be equally great. But the options are open and the choice will have to be made before long as to which path to follow—well-prepared measures to steer its course towards a better society or a *laissez-faire* attitude which will seek to modify policies and conditions to absorb its consequences as they appear. In the immediate future, the major task will be to probe deeper into the subject and try to define more accurately the nature and dimensions of the probable consequences in preparation for decision on the option to follow. For example, if universal application of the technology does create the resources to make possible the "jobless society", deep attention is necessary to discover the best means to distribute the benefits of the new surge of growth. Traditional methods would merely bring the profits to the few and minimum benefits to the frustrated majority. The microelectronics revolution will only be truly revolutionary if it succeeds in creating a society of equity, with a high degree of industrial democracy and the possibility of creative fulfilment to the many. In this, as in other aspects of this complex tangle of promises and doubts, conventional wisdom, status quo policies and linear thinking will not break through to the new society which is intrinsically possible. The development of microelectronics must lead towards new approaches and new patterns of thinking. It is the way either to a mechanised world of alienation and resentment or of the enrichment of individual life and an enhancement of cultural diversity. The choice is ours and will not wait for our successors.

11

Occupation versus Work

ADAM SCHAFF

This is the final chapter of the book, but it has neither the function of summing up its reflections nor is it even a direct consequence of them. This book was consciously limited to the likely events of the next ten years, whereas the really significant breakthrough of microelectronics into the domain of social implications should surely be placed beyond this time limit. The reflections of this chapter go at least two decades beyond the present decade. Does this mean that we are indulging in a sort of speculative futurology or perhaps science fiction?

The answer is neither. In the author group we agreed that while trying to be very modest and sober in our forecasts, we also needed some vision in our reflections about this world which is in the process of transition under the impact of microelectronics on social life. Even if exact data are lacking, this vision is particularly necessary now since it concerns problems crucial for the future development of society. Among such problems, man's essential need for work and occupation assumes a very great importance.

The message in this respect is a very simple one: the microelectronic revolution will undoubtedly change the role of work in human life, diminishing the need for it and in some cases eliminating it totally. This will create a problem of how to replace the traditional "sense of life" of human beings, which is linked in the first place and especially in the industrialised societies of the North with work. The question is how to prevent pathological phenomena in the life of society, especially as far as young people are concerned, if this lacuna is not filled sufficiently early with appropriate content.

338   *Microelectronics and Society*

We can discuss—and hard disputes are going on already in this respect—whether work will be eliminated by the microelectronic revolution in this or that domain, in this or that dimension; whether the new technology will create enough new jobs to replace the loss of the old ones, etc. All these questions remain open. But nobody, knowing the trend, can deny that there will be less need for physical work, that working hours will become shorter and shorter in production and traditional services in the long run—say twenty to thirty years from now, when children born today will be active in society. This prospect is so imminent that a discussion of whether those people should be regarded as "working" or not becomes futile. The social problem of losing the "sense of life" concerns not only those who will not work at all (in the traditional sense of the word) but also those who will work one day or a few hours per week. The problem—a huge problem—remains.

One could certainly say: leave those people alone, they will find their own way of life. But this view can only come from those who understand little or nothing of human psychology. We shall come back to this "argument". During the work on this book I had the feeling that it was obligatory to cope with this problem now—the time is short. If the future shows that the problem is void—we shall forget it easily. But if it proves that we have to do with a real and important social problem, it would be an unforgivable sin to skip the message only for the sake of understatement. And therefore we take the risk of including in the book a bit of "vision", avoiding, however—as it seems to me—speculation and science fiction. It is a forecast founded on a high degree of probability. And this should suffice.

*   *   *

It is obvious that changes in the various aspects of social life caused by the application of microelectronics will look different in the initial period of the process from what will take place when the process is in full swing. But this later period is not very remote: it is a matter of some twenty to thirty years, i.e. the advent of the next generation. The problem must be thought over, especially when it comes to those measures which are intended to prevent the possible and foreseeable adverse effects of the process in the societal field. Compensatory measures must be devised early, because the elaboration of a pro- gramme of appropriate actions and the preparation for its implemen- tation will be time- and labour-consuming. This applies in particular to the fundamental social issue, namely the elimination, to a great extent, of human work in the traditional sense of the word by full automation. This is particularly important, since it means under- mining, if not abolishing, the goal which now guides human actions *en masse.* And it is this goal which programmes and controls human actions through being the essential content of what in an esoteric language is often called "the sense of life". That element of the syn- drome of the effects to be caused by the applications of microelec- tronics in societal life is certainly one of the most difficult to cope with when it comes to its possible adverse consequences. To cope with it successfully we must have both imagination and adequate knowledge, which can be obtained only through a collective and interdisciplinary effort. It is quite clear that what is being said here does not pretend to do anything else than to pose the problem and to make certain sugges- tions concerning the methods of handling it.

Work and Occupation

Before we proceed to submit such suggestions, we have to make cer- tain preliminary remarks intended to impart precision to the terms used in the text and to bring out the problems, even at the cost of repetition of some formulations which are explicitly or implicitly con- tained in the earlier chapters of the book, though in a different form and with a different purpose in view. This applies in particular to the chapters concerned with technological issues linked to microelec- tronics.

As has been said above, full automation will, in the long run, largely eliminate human work *in the traditional sense of the word.* What does this reservation in italics mean and why has it been added? It has been added deliberately and is due to the conviction that even *full* automa- tion of production and services, while eliminating human work, can- not eliminate human activity of various types and kinds. In order to understand the sense of that statement we must define what we mean by *work.* The word *work* means here, in accordance with common understanding, human activity intended to obtain a certain effect in production (production of goods) or in services (satisfaction of certain other human needs) through expenditure of physical or mental energy. It is on this basis that we make the distinction between manual and mental (brain) work. In the case of the societies based on commodity exchange and hired labour, the definition must include the intention of earning money in exchange for being engaged in production or services. All work is activity, since it assumes the purposeful expenditure of human energy, as every activity does. But not all activity is work, because not all activity is undertaken with the intention of earning money by producing goods for the satisfaction of the material needs of human life or implementing some services: it may be creative (artistic, scientific, etc.) activity, or activity in which one engages during one's leisure time, etc.

To avoid verbal misunderstandings and to facilitate further discussion, we stress that full automation will largely eliminate *work* in production and services, but will not put an end to human *activity* and, in this sense, to human *occupation*. It will lead to the replacement of what was previously "work" by creative, entertaining occupation. Thus, our requirement, to be put forward later, that work should be replaced by other forms of human activity (occupation), which will restore to man his goal in life, or, in other words, his "sense of life", is quite realistic.

In Which Domains will Occupation Remain Non-automated?

If we think realistically, we have to pose the question: will *all* people in the long run lose, as a result of the expected course of events (i.e. full automation), the possibility of being engaged in some useful and attractive work? The answer is of course not. Let us accordingly examine more closely the prospects in order to be able to answer the question better.

There is no doubt that in spite of all the advances in automation it will primarily cover the production of material goods and partly services, including the various kinds of clerical work. But all kinds of socially useful "higher" activity and various organisational functions will remain, and in some cases will even be greatly expanded. Thus a large part of society will still find employment, but within a changed employment structure: these changes will affect the traditional branches and there will be new fields emerging and older ones expanding according to the new needs of society and its greater wealth. In other words: the traditional working class will diminish or vanish (as was foreseen by Marx in the mid-nineteenth century) as an effect of automation. The same will, to a large extent, apply to farm labourers, clerks, and a considerable part of the people now employed in the services. But the various traditional fields, especially those connected with intellectual functions, will remain, and in some cases will absorb increased numbers of people. New fields will also emerge in which large parts, if not the majority, of the population will find employment. It is difficult to give numerical estimates, even within the framework of a risky futurology, but a cursory glance at the list of the surviving, expanding and newly emerging spheres of human activity must calm the minds of those disturbed by the prospects of structural unemployment. Here is that list, certainly far from being complete, with its items selected only by way of example.

The first group of occupations which will certainly remain and witness great expansion, both because of the growing needs in that respect and because of the growing population of those who will get access to this domain of activity, includes all the spheres of *creative* work (occupation). Pride of place here goes to research, which will increasingly become the socially most important tool of production. Next there will be art in all its forms and manifestations, whose popularisation for the masses will increase the number of people engaged in it (note that this field covers the cinema, TV, radio, etc.). Further, there will be architecture and applied art to embellish the home, and other artistic crafts, e.g. fashion design, and so on.

The second vast and ramified sphere of human occupation which will not be automated, even though like the former it will be widely assisted by automata of all kinds (above all by computers), is the organisation of societal life. Even if decentralised in some of its aspects, as is already postulated, it will of necessity tend toward a comprehensive approach, especially in the sphere of economics, on the regional and even on the global scale. This covers the planning and managing of the various social phenomena; the study of public needs and their trends; satisfaction of public needs by the expansion of the network of shops and banks; health care; legal aid; schools of various types and levels; transportation; various types of environmental conservation; restaurants and hotels; police force; etc.

The third sphere, largely innovative and becoming possible due to increasing social wealth and a larger supply of the appropriate manpower, will include the greatly expandable network of social consultants in the form of social workers, to be concerned especially with old people, with disabled people and with invalids of all kinds, with young people seeking advice about their future activities, about specialised continuous studies (other than those provided by the established curricula), about family planning, etc.

The fourth sphere will include production and services, and inspection in these fields. Here there will be highly trained or skilled technicians who will replace the working class, and certain types of clerical staff. Let us not forget the need for the maintenance of equipment and a whole army of people dealing with problems of automation.

The fifth sphere will cover the organisation of leisure time, primarily the various sports (available to all) with a large number of instructors, tourism, various cultural institutes.

Consequently, there will remain a huge realm of occupations covering the domain of classical work. Nevertheless, automation will create problems.

Microelectronics and Automation

Our second remark refers to the problem of whether the development of microelectronics and its applications in societal life really lead inevitably to the automation of production and services, thus gradually eliminating human work in the corresponding sectors. As we know from what has been said in the preceding chapters, the answer to this question is in the affirmative. Automation of a given sphere of production or services does drive out human work, resulting thereby in a structural unemployment in that sphere, not because of any transient oscillations in business trends, but because of its permanent effects through the production of goods and performance of services by automata. That tendency will intensify universally as new spheres of societal life become covered by automation.

For the time being we can stop at that, in the hope that the analyses made in the previous chapters have clarified the issue: the development of applications of microelectronics in production and services makes automation, including its ultimate (i.e. full) form, a necessity. This cannot be helped, and there is moreover no need to counteract that trend because it has manifold and already now manifest positive effects in different realms of human life. The process cannot, in consequence, be stopped now. If this is really so, then we should diligently concern ourselves with the societal consequences of that process, especially when it comes to the danger of growing structural unemployment.

There are two categories of people who oppose this statement. The first includes the advocates of a specific policy of *laissez-faire.* They argue roughly as follows: we should not be troubled by those prospects because often in the past, when production techniques changed radically, mankind faced a similar danger and managed to cope with it successfully. New branches of production and services always emerged and expanded, and new jobs were offered which not only absorbed the unemployed but also created demand for new workers. They take as an example the effects of the first industrial revolution.

This is erroneous reasoning which may also have dangerous consequences. The fact that in the past the consequences of such crises were remedied spontaneously does not mean that it will be so this time too. The reasoning is burdened with a logical *non sequitur,* because the quite different nature of the present process not only affects the existing branches of production and services, but also influences the future ones. People who will be put out of their jobs by automation must be offered occupation of another kind and a chance of activity of another type. The triumphant dismissal of the problem, which is often due to the fear of facing it or even is a deliberate lie intended to protect the present-day interests of certain business groups, merely demoralises society and dissuades it from engaging in necessary analyses and preparations. And this is dangerous.

That is, the first category of opponents of our view tends to dismiss the problem, at least apparently.

The second category includes the genuine advocates of the policy of *laissez-faire,* who programmatically proclaim that the matter should be left to its spontaneous development and should not be interfered with by any attempts to control the process. The advocates of this opinion argue as follows: there have been primitive societies in which the time free from productive occupations (hunting, food gathering, etc.) was spent spontaneously on play and rest, and there was no problem with it. Modern man should accordingly be left to himself, with his leisure time, and he will find ways of spending it. They even claim that this is a *concrete* approach to the problem.

First, men in less developed societies, both in the past and today, toil hard to meet their elementary needs, and it is only the rest of their time, free from toil, which they fully enjoy. Whereas on the contrary, contemporary man in highly industrialised societies faces the danger of a "pollution" of his leisure time if he is freed from the need to work and is left with his leisure time only. His leisure will not mark an interval between periods of work, but will cover all his life. And he also faces the danger of the impact of automation on his activity, his thinking and way of life.

Secondly, which is the most important point, contemporary man—if we formulate the problem in *concrete* terms and not speculatively—has other needs than primitive man had before, and anyone who fails to understand this factor of the "culture code" in human evolution does not comprehend anything of the problem we face. One of the issues involved is the social status which the individual wants to enjoy and the social role he wants to play in contemporary society, and this is linked to the cultural patterns shaped by human history. Occupation is needed to supply life goals. The point is—and it is particularly important in the case of young people—that the individual should become independent within the framework of society and find his place and role in it. It is pointless to refer to simpler communities of the past and to advocate a *backward* movement in the history of mankind. Except as a result of great disasters, nothing moves backward in history. We have to move forward, that is from work to new, higher forms of human activity. If work currently constitutes man's life goal and forms the context in which he puts into effect his aspirations to become independent within society, to enjoy the adequate social status and to play the adequate social role, then, when work vanishes as a result of the application of new technologies, we have to find a new way and a new context for the attainment of his goals. And we have to do that on a *societal* basis, and not to leave it to the individual himself. In other words, the society will have to constitute new goals in human life as possible alternatives from which the individual may choose and hence achieve "the sense" of his life. This process has always been, and will continue to be, a *societal* one. The problem narrows down to whether that process is to be spontaneous or consciously guided by societal reflections and organisation.

Of course, I declare myself for the second solution. Allowing the process of adjustment to be spontaneous may cost us too much and may result in deep-reaching social pathology. Those who suffer from a frustration caused by an excess of authority and bureaucracy in contemporary societies should realise that the point is not to impose upon anyone any new goal in his life, but—in view of the fact that the old traditional goal is vanishing—to offer him alternative solutions and to outline new possibilities, leaving it to every mature individual to choose one of those various options. Those who advocate spontaneity should notice the dangers which their proposal entails.

To put it in concrete terms, if in forty to fifty years from now we have tens, if not hundreds, of millions of *structurally* totally or partly unemployed people in the industrialised countries, they will include women, the elderly, the disabled and predominantly *young* people. To rely on spontaneous adjustment to provide for those millions of people, especially young people, means to doom them to frustration, social pathology and rebellion. This tendency has already found expression in drug addiction, alcoholism, violence and delinquency. We have to realise that these young people, with the exception of those who are highly educated and prepared for special jobs, will find the world closed to them. Their opportunities for implementing the received patterns of life will be ruined. If society does not offer them a *real* alternative, it will doom them to pathology. The point is to find measures which will avert that danger.

When talking about unemployed people in this future society, we must be fully aware of the change in connotation of the word "unemployed". This society will be not only a welfare society, but the distribution of wealth, its social structure and the social mobility of its members will also be deeply changed. From the point of view of its social system, pluralistic solutions, due to different causes, must be envisaged. But in all of them society will have to take over the implementation of the needs of its members, on a historically given level of these needs, specifically because of the existence of a structural unemployment. The society could not survive otherwise. But this means that the unemployed person, unlike now, will have all his material needs covered and will not suffer materially from his unemployment. The only thing lacking will be work. Taking into account, however, the human need for an active life and the problem of social status linked with work—especially as far as young persons are concerned—this deficiency would be a serious predicament for mankind.

Continuous Education as a Form of Universal Occupation

As we have seen, the list of future non-automated occupations is quite imposing, and more so as—let this be repeated once more—it is far from being complete. Quite understandably, the number of those formally employed in a given field will often be minimal as compared with today, because people will be assisted by improved machines and automata. But it is equally true that in many fields the scope of activity will be broadened greatly, which will increase the number of those employed or involved.

As a further element we should also take into account a new way of distributing work, i.e. providing work opportunity for more people with reduced working hours. There is, however, a limit to this process beyond which (e.g. working one day or only a few hours a week) it loses its sense—man would find himself psychologically in the same situation as if he did not work at all.

In addition, various activities can be greatly extended on an amateur basis. Support by the state or local authorities in the form of providing training facilities or tools could greatly facilitate this. This possibility would apply to a broad spectrum of activity—artistic, scientific, craft, sport, etc. One can envisage, for example, that the average person in the future will spend a relatively important part of his life in one or several subsidiary occupations which generate satisfaction and self-fulfilment. Thus the picture does not look as pessimistic as it might seem at first: not all people will be deprived by automation of the opportunities to be active, and they will, moreover, have many options in that respect. But one thing is alarming: in many cases these spheres of action will require high, and even very high, qualifications. Hence, not everyone will meet the requirements and, moreover, there will be, on the global scale, millions of people for whom there will be little or no work in the traditional sense of the word in view of the almost complete disappearance of the working class and white-collar groups. Surely we can assume a considerable shortening of working hours that would add new jobs, but as we have already said, it will not eliminate the "pollution of leisure". What is to be done with that part of the population, how can we find occupation for all? The simplest solution would consist in introducing continuous education in diversified forms and alternative models, alternating with traditional work or other occupations, to be elaborated for all people until retirement.

This would settle two matters of great social significance. *All* people would find a sensible field of activity and hence a new "sense of life", and society at large would have an opportunity of putting into effect the old humanistic ideal of developing "a universal man", i.e. one who has many-sided education and is thus capable of changing his occupation as need may be. The requirement of continuous education would be made realistic by the fact that people who are now regarded as structurally unemployed would receive from society means of subsistence at the level of needs shaped in the process of history. Society would thus be entitled to demand of them, as part of their responsibility to it, certain reciprocal performances which would be made obligatory—during a certain period of life—for all. This would be like the compulsory education of young people which has already become a fact in many countries.

The idea of continuous education implies the necessity of modifying the *entire* school curriculum. This is a task for educationalists who will have to take into consideration the needs and the specific cultural features of the various regions and countries. The comments that follow are merely intended to contribute to the discussion.

Regardless of all the possible changes in the educational system, there is no doubt that a certain period of general training should be maintained in order to transmit to the younger generations general knowledge and culture and an adequate system of methods of mental work as an introduction to the higher levels of education and to provide the tools to enable the learning process to continue throughout life. If we consider the goal of that general training and the needs and possibilities of society, we have to conclude that this period would be much longer than is obligatory today. Since such general education should include disciplines which are now neglected or even abandoned, such as, for instance, aesthetics with an introduction to the various spheres of artistic work, rhetorics as a preparation for active participation in public life, etc., and there should also be an introduction to information technology, it is not an exaggeration to claim that such general education should continue longer than now.

Of course, the curriculum and the system of schooling would have to be reformed and diversified and should be by no means exclusively academic. The curriculum would have to be adjusted to the social and cultural requirements of a given region and/or country. The student's memory should not be taxed too much, especially in view of the opportunities offered in this respect by computer memory. And, in the later years, the students would have to enjoy more freedom to individualise their process of learning, in line with the patterns that have for a long time been used in the various national educational systems. To keep this system working it must differ radically from the traditional one: there must be diversification of types of learning; the linkage of theoretical studies with practical, also manual, work; opening the way to individual options linked with the gifts of the student, etc.

The second stage of education (where reforms are also needed) should combine studies in a selected discipline with activity as an instructor in a given sphere of public education or in some other field. The point is that on completing this universally obligatory period of general education, a person should engage in active societal life, combining it with further studies, which would ensure his greater maturity as a member of society and as a student at a higher level. According to the selected course of study, this second period should continue until one is twenty-five to thirty years old. The nature and amount of his work as an instructor should be a function of the requirement of his second-level studies.

The third stage would be that of employment, if the person meets the appropriate requirements. And there would be a continuation of studies in freely selected fields, combined with occupation as a higher-level instructor, if the person in question does not become permanently employed.

It is self-evident that the appropriate groups of experts would have to work out the curricula of such facultative continuation studies. These could consist in a person's deepening his knowledge in his main field of study, or in his studying other subjects in accordance with his interests. Different models of combining employment and further studies should be elaborated and implemented.

What would be the advantages of such a programme?

First, a radical and socially useful solution of the problem of structural unemployment.

Second, the implementation of the principle of permanent social dynamics: increasing the level of societal life and striving for the pattern of "universal man" for everyone.

Third, a radical change in the nature of society whose members would pass from the stage of *Homo laborans* to that of *Homo studiosus,* without losing the nature of *Homo ludens.* That would mean unquestionable progress.

All this, when seen from today's point of view, looks like Utopia. But we have to bear in mind, and be imaginative enough to see, that a change in perspective has often changed Utopia into reality. The present reflections, preliminary and sketchy as they are, bring out one point: on our path to the implementation of such plans we will have to do immense conceptual and organising work on a global scale. If we consider the fact that what is now being planned will become a necessity in some thirty years, we must conclude that it is high time to engage in the appropriate preparations. In practice, we are already lagging behind the requirements of our time, and such a delay may be both costly and dangerous to society.

Glossary of Technical Terms

I

Algorithm: Sequence of steps required to perform a calculation or a logical process.

Assembly language: A low-level symbolic programming language usually expressed as combinations of two or three easily memorised letters.

Baud: A numerical system based on two digits, "1" and "0".

Bit: A unit of information—the abbreviation of binary digit, one of two values used in binary notation (1, 0).

Boolean algebra: A branch of mathematics based on the work of George Boole, in which logical propositions are denoted by symbols and resolved with binary operations.

Bubble memories: Semiconductors containing a film of charged moving bubbles used for storing binary data.

Byte: An eight-bit binary number.

CAD: Computer assisted/aided design.

CAM: Computer assisted/aided manufacturing.

CCD: Charge-couple device, a solid state memory chip which stores information in the form of packets of electrical charges.

CCM: Computer coupled machines used in the computer assisted manufacturing process.

Chip: A square or rectangular piece of crystal, about 0.5 cm per side, usually of silicon or germanium, which contains an integrated circuit.

Components: Minute electronic devices making up part of an integrated circuit and generally considered to be of two types, active (transistors) or passive (resistors, capacitors).

CPU: The central processing unit of a computer, where logic and arithmetic functions are performed.

EPROM: Erasable programmable read-only memory.

Floppy disk: Flexible ferrite disks used for the storage and retrieval of information for use in a computer.

Formants: Units which represent sounds in terms of their resonant frequencies and amplitudes.

High-level language: Language close to the language of the user for communicating with a computer.

Informatique: The science by which computers store and process information.

Integrated circuit: A semiconductor chip holding a circuit formed of transistors, resistors, diodes, etc.

Languages: The means by which the user communicates with the computer.

Lasers: An acronym for light amplification by stimulated emission of radiation; these produce coherent light, that is where the atoms radiate in step with one another.

Logic gates: Electronic circuit elements which are open or closed depending on whether propositions are true or false.

Machine language: Combinations of binary digits which the computer understands.

Microcomputer: A printed circuit board containing a microprocessor, RAM and/or ROM memory and input and output devices. Also sometimes a computer on a chip.

Microprocessor: An integrated circuit on a chip where the logic and arithmetic functions can be performed.

NCM: Numerically controlled machine tool.

Phonemes: A set of frequencies expressed electronically by the binary system used to represent basic speech sounds.

PROM: Programmable read-only memory.

RAM: Random access memory. A flexible, magnetised disk sensitive to erasure which gives instant access to information independent of sequence in which it was stored.

ROM: Read-only memory, containing permanently stored information.

Semiconductor: An electronic conducting material such as silicon in which the resistance to the flow of electrons is between that of metals and insulators.

Software: The operating instruction provided to the computer. It includes programming languages and machine codes.

Transistor: A solid state tiny electronic device consisting of a piece of semiconducting material such as silicon.

Videotex: An interactive visual communications system which accesses stored information through telecommunications links.

Wafer: A thin slice of a semiconducting material such as silicon, several centimetres in diameter, which contains several hundreds of identical integrated circuits.

II*

Air-launched cruise missile (ALCM): A cruise missile designed to be launched from an aircraft.

Anti-submarine warfare (ASW): Warfare against submarines, involves: (1) detection; (2) classification (to distinguish friendly from hostile submarines); (3) localisation; (4) attack and (5) destruction.

Assured destruction: The ability to inflict an "unacceptable" degree of damage upon an aggressor after absorbing any first strike. Mutual assured destruction is a condition in which an assured destruction capability is possessed by opposing sides.

Ballistic missile: Any missile which does not rely upon aerodynamic surfaces to produce lift and consequently follows a ballistic trajectory (i.e. that resulting when the body is acted upon only by gravity and aerodynamic drag) when thrust is terminated.

Circular error probable (CEP): A measure of the delivery accuracy of a weapon system used as a factor in determining probable damage to targets. It is the radius of a circle around the target at which a missile is aimed within which the warhead has a 50 per cent chance of falling.

*Based on a United States Arms Control and Disarmament Agency Lexicon.

Cruise missile: A guided missile which uses aerodynamic lift to offset gravity and propulsion to counteract drag. The major portion of a cruise missile's flight path remains within the earth's atmosphere.

Deterrence: Any strategy whose goal is to dissuade an opponent from attacking.

Electronic counter-measures (ECM): Electronic warfare (EW) involving actions taken to prevent or reduce an enemy's effective use of equipment and tactics employing or affected by electromagnetic radiations and to exploit the enemy's use of such radiations.

First strike (nuclear): The launching of an initial strategic nuclear attack before the opponent has used any strategic weapons himself.

Intercontinental ballistic missile (ICBM): A land-based, rocket-propelled vehicle capable of delivering a warhead to intercontinental ranges (ranges in excess of about 3000 nautical miles).

Inertial guidance system: A guidance system designed to project a missile to a predetermined point on the earth's surface by measuring acceleration.

Launch-on-warning: A doctrine calling for the launch of ballistic missiles when a missile attack against them is detected and before the attacking warheads reach their targets.

Mach: The speed of sound, e.g. Mach 4 means 5 times the speed of sound.

Second strike: A term usually used to refer to a retaliatory attack in response to a first strike.

Sensors: Devices used to detect objects or environmental conditions.

Strategic arms limitation talks (SALT): A series of negotiations between the United States and the Soviet Union which began in Helsinki in November 1969. The negotiations seek to limit and eventually reduce both offensive and defensive strategic arms.

Strategic nuclear weapons systems: Offensive nuclear weapons systems designed to be employed against enemy targets with the purpose of effecting the destruction of the enemy's political/economic/military capacity and defensive nuclear weapons systems designed to counteract those systems.

Submarine-launched ballistic missile (SLBM): A ballistic missile carried in and launched from a submarine.

Tactical: Generally, relating to battlefield operations.

Warhead: That part of a missile, projectile, or torpedo that contains the explosive intended to inflict damage.

Yield: The force of a nuclear explosion expressed in terms of the number of tons of TNT that would have to be exploded to produce the same energy.